HANDBOOK OF APPLIED MICROBIOLOGY

HANDBOOK OF APPLIED MICROBIOLOGY

Edited by

Dr. Dnyaneshwar M. Pawar

Delve PUBLISHING

www.delvepublishing.com

Handbook of Applied Microbiology

Dr. Dnyaneshwar M. Pawar

Delve Publishing
4164 Lakeshore Road
Burlington, ON L7L 1A4
Canada
www.delvepublishing.com

Email: orders@arclereducation.com

ISBN: 978-1-77956-363-7 (Hardcover)

ABOUT THE EDITOR

Dr. Dnyaneshwar M. Pawar (1985) is presently serving as Assistant Professor, Department of Plant Pathology, College of Agriculture, Naigaon (VNMKV, Parbhani) Maharashtra. He obtained his B. Sc. (Ag.) in 2008 from Marathwada Agriculture University, Parbhani. M. Sc. (Ag.) 2010 & Ph.D. 2014 in Department of Plant Pathology, Navsari Agriculture University, Navsari (Gujarat). He started his career as Assistant professor in 2014. His field of specialization is bacteriology, mycology and virology. He has edited 02 books in international level, published 13 research paper in National and international journals of reputed. He has participated many National, State Seminar and symposiums. He has written 01 practical manuals for under graduate course for the benefit of students besides several Radio talk, Lectures in farmer training programme. Dr. Dnyaneshwar Pawar has also written 2 popular articles in reputed magazines.

TABLE OF CONTENTS

LIST OF FIGURES

ABSTRACT

The Handbook of Applied Microbiology is a comprehensive guide that offers practical information on the applications of microbiology. This resource covers a wide range of topics, including microbial biotechnology, environmental microbiology, and industrial microbiology, making it highly valuable for researchers and professionals in the field. The book holds great significance due to the growing importance of microbiology in modern society. Microorganisms are now widely utilized in various biotechnological processes, environmental monitoring, and food production. Having a deep understanding of applied microbiology is essential for ensuring the success and safety of these applications. *The Handbook of Applied Microbiology* provides in-depth insights into cutting-edge research and technologies, empowering readers with the knowledge and tools necessary to tackle present-day challenges and seize opportunities in the field. This resource caters to a wide range of individuals, including students, researchers, and industry professionals. Overall, the book is a valuable reference for advancing research and innovation in applied microbiology.

PREFACE

The Handbook of Applied Microbiology covers a broad spectrum of subjects concerning the practical utilization of microbiology across diverse domains. This book encompasses the most recent advancements, methodologies, and discoveries in the field of microbiology, ranging from environmental to industrial applications. The text offers a thorough examination of how microbiology is applied to address practical challenges and enhance operations across multiple sectors.

The significance of this work lies in its ability to bridge the gap between academic knowledge and practical applications in the field of microbiology. By presenting tangible examples and empirical analyses, this publication helps readers comprehend the importance of microbiology in daily life and its potential role in addressing urgent concerns such as ecological contamination, food security, and disease mitigation. *The Handbook of Applied Microbiology* is an invaluable resource for students, researchers, and professionals seeking to apply microbiological principles in their respective fields.

The material presented in this book is essential, given the increasing importance of microbiology across various industries and disciplines. Issues like antibiotic resistance, climate change, and emerging infectious diseases underscore the importance of microbiology in tackling these challenges. This book equips readers with the knowledge and tools necessary to apply microbiology effectively for positive outcomes in their specific areas.

The Handbook of Applied Microbiology stands out due to its comprehensive coverage of a wide range of topics and its pragmatic approach to microbiology, making it a valuable read. Unlike traditional microbiology textbooks, this book emphasizes the practical applications of microbiology and its ability to solve complex issues. Through advanced research and innovative solutions, it encourages readers to think creatively and apply microbiological principles in novel and impactful ways.

This book will appeal to a diverse audience, including students, researchers, educators, and professionals in microbiology, biotechnology, environmental science, food science, and healthcare. Whether you are a student exploring career opportunities in microbiology, a researcher aiming to stay current with recent advancements, or a professional seeking to expand your expertise, this book provides valuable insights for all.

Readers will benefit greatly from this book. It offers a deeper understanding of the practical applications of microbiology and its potential to address real-world challenges. The book encourages critical and innovative thinking about the practical uses of microbiology in professional settings. Additionally, it provides significant insights into current trends and technology in microbiology, enabling readers to remain at the forefront and make informed decisions in their respective fields.

Overall, *The Handbook of Applied Microbiology* is an invaluable reference for anyone interested in exploring the practical applications of microbiology across various industries and fields. Offering an in-depth examination of the latest research, methods, and advancements, this book empowers readers with the knowledge and tools to apply microbiology effectively for positive outcomes. It serves a wide audience, including students, researchers, educators, and professionals, by delivering valuable content for all. Explore the world of applied microbiology and uncover the boundless opportunities ahead.

CHAPTER 1

INTRODUCTION TO MICROBIOLOGY

Contents

Microbiology, a fundamental branch of biological sciences, studies microorganisms such as bacteria, viruses, fungi, protozoa, and algae. This scientific discipline is critical for understanding the complex roles microorganisms play in a variety of ecological, industrial, and medical contexts. Microbiology began with the pioneering work of Antonie van Leeuwenhoek in the 17th century, whose meticulous observations of microscopic entities laid the groundwork for modern microbiology (Murray et al., 2015). Bacteriology, virology, mycology, and parasitology are all subfields of modern microbiology that contribute to a comprehensive understanding of microbial physiology, genetics, and ecology.

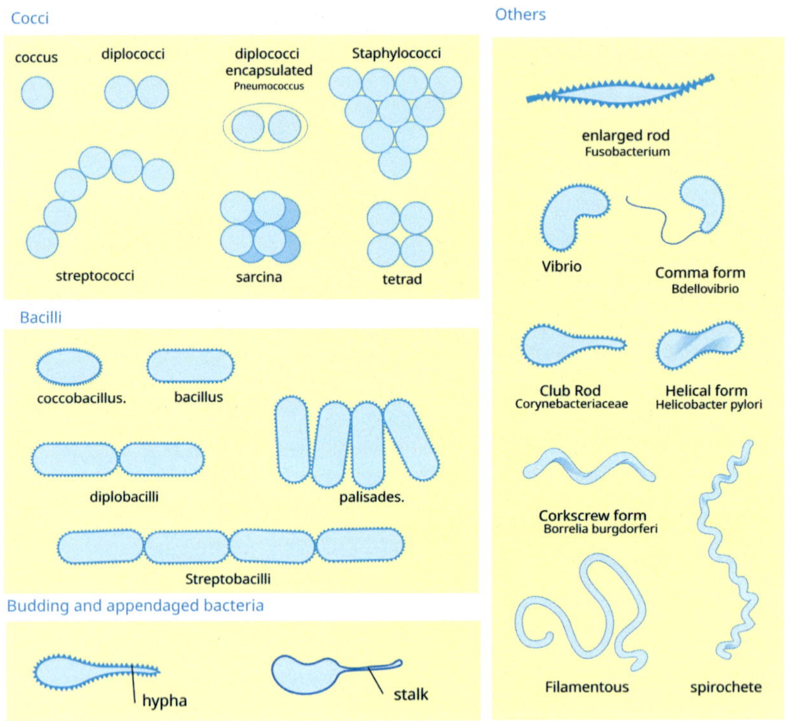

Figure 1.1. Microorganisms.

Source: By Mariana Ruiz LadyofHats – Own work using: Adobe Illustrator +[1][2][3][4][5][6][7]Medizinische Mikrobiologie, Ernst Wiesmann ED. Thieme (1986), Public Domain, https://commons.wikimedia.org/w/index.php?curid=738916.

The development of molecular biology techniques has transformed microbiology, enabling advances in microbial genomics, metagenomics, and synthetic biology. These innovations have shed light on microorganisms' genetic and metabolic diversity, revealing their critical roles in nutrient cycling, bioremediation, and pathogenesis. Furthermore, microbiology intersects with public health due to its importance in the study of infectious diseases, antibiotic resistance, and the development of new therapeutic strategies (Maier et al., 2009). Microbiology remains an indispensable field, driving scientific progress and fostering a deeper understanding of the microscopic world, which has a profound impact on life on Earth.

1.1. HISTORY OF MICROBIOLOGY

Significant milestones in microbiology's history have shaped our understanding of the microbial world as a whole. This journey began in the 17th century with the pioneering work of Antonie van Leeuwenhoek, a Dutch tradesman who, using handcrafted microscopes, was the first to observe and document microorganisms, which he referred to as "animalcules" (Murray et al., 2015). His meticulous observations laid the groundwork for the study of microbiology.

Louis Pasteur and Robert Koch, widely regarded as the fathers of microbiology, made transformative contributions in the 19th century. Pasteur's ingenious swan-neck flask experiments in the 1850s and 1860s disproved the theory of spontaneous generation by demonstrating that microorganisms exist in the air and can contaminate sterile solutions (Murray et al., 2015). He also developed pasteurization and made significant contributions to the study of fermentation and vaccination, particularly against rabies and anthrax.

In the late 19th century, Robert Koch revolutionized microbiology by developing Koch's postulates, a set of criteria for determining a causal relationship between a microbe and a disease. His research on the causes of tuberculosis, cholera, and anthrax established a solid foundation for medical microbiology and the study of infectious diseases. Koch's techniques for isolating and cultivating bacteria were pivotal in the development of pure culture techniques and solid media (Gest, 2004).

The beginning of the twentieth century heralded the golden age of microbiology, marked by rapid discoveries and technological advancements. Hans Christian Gram developed the Gram stain in 1884, which became an important method for bacterial classification and diagnosis. Alexander Fleming's discovery of penicillin in 1928 signaled the start of the antibiotic era, fundamentally changing the treatment of bacterial infections (Maier et al., 2009). This period also saw the emergence of virology as a distinct discipline, with Wendell Stanley isolating the tobacco mosaic virus in 1935, demonstrating that viruses differ from other microorganisms due to their unique structural and functional properties (Gest, 2004; Murray et al., 2015).

With the introduction of genomics and metagenomics in recent decades, the field of microbiology has expanded into new frontiers, allowing for the comprehensive analysis of microbial communities and their functions in a variety of environments. The Human Microbiome Project, which began in 2007, has provided detailed insights into the complex interactions between human health and the microbiota, revealing microorganisms' profound influence on human physiology and disease (Murray et al., 2015). Another burgeoning field, synthetic biology, seeks to engineer microorganisms for a variety of applications, including biofuel production and environmental remediation.

1.2. MICROBIAL CLASSIFICATION

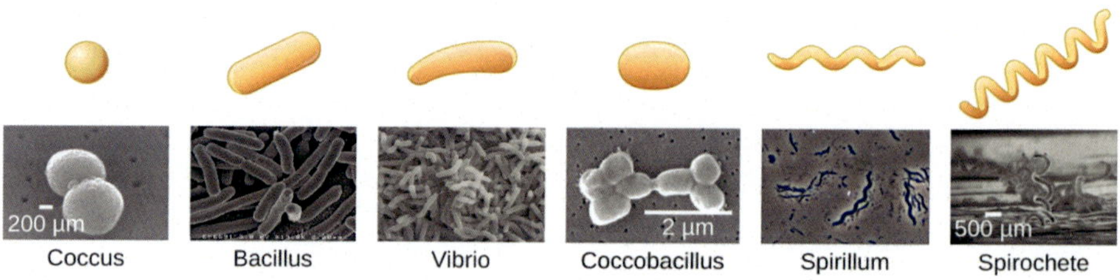

Figure 1.2. *Classification of microorganisms.*

Source: https://courses.lumenlearning.com/suny-microbiology/chapter/types-of-microorganisms/.

Microorganism classification is a fundamental aspect of microbiology, providing a systematic framework for identifying, naming, and categorizing the wide range of microscopic life forms. This classification is based on taxonomy, which includes hierarchical levels such as domain, kingdom, phylum, class, order, family, genus, and species (Maier et al., 2009). The modern classification system combines molecular and genetic analyses, increasing our understanding of microbial diversity and relationships. Bacteria, archaea, fungi, protozoa, algae, and viruses are important groups of microorganisms, each with its own characteristics and classification criteria (Madigan et al., 2018).

1.2.1. Bacteria

Bacteria are prokaryotic microorganisms that lack nuclei and membrane-bound organelles. They are classified as Bacteria and show remarkable diversity in shape, size, metabolic pathways, and habitats. Traditional classification was based on phenotypic characteristics like morphology, staining properties (e.g., gram stain), and biochemical tests (Madigan et al., 2018). However, molecular techniques, particularly 16S ribosomal RNA gene sequencing, have transformed bacterial taxonomy, enabling more precise phylogenetic relationships.

Bacteria are classified into several phyla, with Proteobacteria, Firmicutes, Actinobacteria, and Bacteroidetes being among the most common. Escherichia, Salmonella, and Rhizobium are among the many genera of proteobacteria that are important in medicine and the environment. Firmicutes are characterized by Gram-positive cell walls and include genera such as *Bacillus* and *Clostridium*. Actinobacteria, which are also Gram-positive, are known for their filamentous forms and antibiotic production, as demonstrated by the genus Streptomyces (Maier et al., 2009).

1.2.2. Archaea

Archaea, which were previously considered bacteria, are now recognized as a separate domain due to their distinct genetic and biochemical characteristics. They are prokaryotic, but differ from bacteria in several ways, including membrane lipids and the presence of unique ribosomal RNA sequences. Archaea are classified into three phyla: *Euryarchaeota*, *Crenarchaeota*, and *Thaumarchaeota* (Madigan et al., 2018).

Euryarchaeota includes methanogens, which produce methane as a metabolic byproduct, as well as halophiles, which live in salty environments. *Crenarchaeota* are frequently found in extreme environments, such as hot springs and hydrothermal vents, as demonstrated by the genus Sulfolobus. *Thaumarchaeota* are known for their role in ammonia oxidation, which contributes significantly to the global nitrogen cycle (Maier et al., 2009; Murray et al., 2015).

1.2.3. Fungi

Fungi are eukaryotic microorganisms that belong to the kingdom Fungi. They come in a variety of forms, including unicellular yeasts, multicellular molds, and mushrooms. Fungi are heterotrophic, acquiring nutrients through absorption and playing critical roles in decomposition and nutrient cycling (Murray et al., 2015).

Fungi are classified into several phyla, including *Ascomycota*, *Basidiomycota*, and *Zygomycota*. *Ascomycota*, or sac fungi, produce spores in sac-like structures known as asci and include genera such as *Saccharomyces* and *Aspergillus*. *Basidiomycota*, or club fungi, produce spores on club-shaped structures known as basidia. Notable members include Agaricus (common mushrooms) and *Cryptococcus*. *Zygomycota*, though less well defined in modern classifications, includes bread molds, such as Rhizopus (Madigan et al., 2018).

1.2.4. Protozoa

Protozoa are a diverse group of unicellular eukaryotic microorganisms traditionally classified as Protista. They have various modes of locomotion, such as flagella, cilia, and pseudopodia, and are frequently classified according to these characteristics (Maier et al., 2015; Murray et al., 2015).

Protozoa are classified into four major groups: Amoebozoa, flagellates, ciliates, and apicomplexa. Amoebozoa, like Amoebas, move with pseudopodia. Flagellates, including *Trypanosoma* and *Giardia*, use flagella to move. Ciliates, including *Paramecium*, have numerous cilia. *Apicomplexa*, including *Plasmodium* (the cause of malaria), are obligate intracellular parasites with complex life cycles (Madigan et al., 2018).

1.2.5. Algae

Algae are photosynthetic eukaryotic microorganisms found in a variety of aquatic habitats. They are classified as Protista and come in a variety of forms, including unicellular organisms and multicellular seaweeds (Maier et al., 2009).

Algae are classified into several major groups based on pigmentation, storage products, and cell wall composition. The major groups are Chlorophyta (green algae), Rhodophyta (red algae), and Phaeophyta (brown algae). Chlorophyta, which includes Chlamydomonas and Volvox, are primarily freshwater algae. Rhodophyta, including Porphyra, are primarily marine and contribute to reef formation. Phaeophyta, which include kelps such as Laminaria, are also found in the ocean and play an important role in coastal ecosystems.

1.2.6. Viruses

Viruses are acellular entities made up of nucleic acids (DNA or RNA) encased in a protein coat. They are obligate intracellular parasites, which means they can only reproduce within host cells. Viruses are classified according to their genetic makeup, replication strategy, and morphology (Madigan et al., 2018).

Viruses are classified using a hierarchical system developed by the International Committee on Taxonomy of Viruses (ICTV), which includes orders, families, genera, and species. The major viral families are Herpesviridae (e.g., herpes simplex virus), Retroviridae (e.g., HIV), and Orthomyxoviridae (e.g., influenza virus). The Baltimore classification system divides viruses into seven categories based on their genome type and replication mechanism, which ranges from double-stranded DNA viruses to single-stranded RNA viruses with reverse transcription (Murray et al., 2015).

1.3. MICROBIAL STRUCTURE AND FUNCTION

Microorganisms have a remarkable diversity in structure and function, which explains their widespread presence and ecological significance. Understanding the structural and functional properties of microorganisms such as bacteria, archaea, fungi, protozoa, algae, and viruses is critical to the study of microbiology. This understanding not only clarifies their roles in various environments, but it also influences medical, industrial, and biotechnological applications.

1.3.1. Bacterial Structure and Function

Bacteria, as prokaryotic organisms, possess a relatively simple structure lacking a nucleus and membrane-bound organelles. The bacterial cell is encased in a cell envelope comprising a plasma membrane, cell wall, and, in some species, an outer membrane. The plasma membrane, composed of a phospholipid bilayer, functions as a selective barrier regulating the influx and efflux of substances (Madigan et al., 2018).

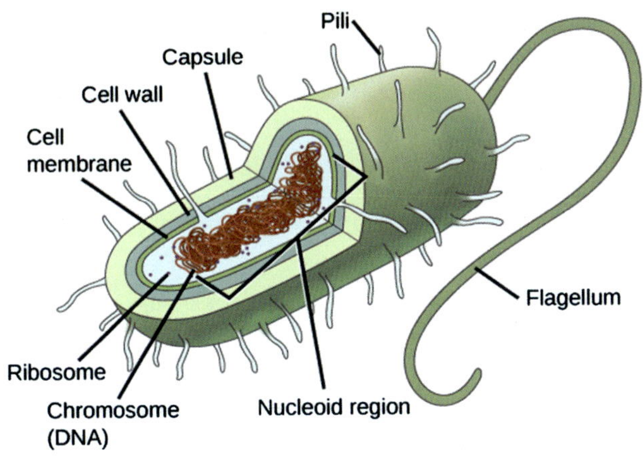

Figure 1.3. Bacterial cell structure.

Source: https://courses.lumenlearning.com/suny-wmopen-biology2/chapter/the-structure-of-prokaryotes/.

The bacterial cell wall, which is primarily made of peptidoglycan, provides structural support and shape. Bacteria are classified as gram-positive or gram-negative according to their cell wall structure. Gram-positive bacteria have a thick peptidoglycan layer, while Gram-negative bacteria have a thin peptidoglycan layer and an outer membrane rich in lipopolysaccharides (LPS). This structural distinction has important implications for antibiotic susceptibility and pathogenicity (Murray et al., 2015).

Bacteria also have various appendages, including flagella, pili, and fimbriae. Flagella are whip-like structures that aid in motility, allowing bacteria to find favorable environments. Pili and fimbriae are hair-like projections that aid in surface adhesion and genetic exchange through conjugation (Maier et al., 2015).

The bacterial cytoplasm contains ribosomes, the nucleoid region that houses the bacterial chromosome, and plasmids. Protein synthesis takes place on ribosomes, which are made up of RNA and proteins. The nucleoid, an irregularly shaped region, houses the bacterial DNA, which is usually a single, circular chromosome. Plasmids are small extrachromosomal DNA molecules that can confer beneficial characteristics like antibiotic resistance (Madigan et al., 2018).

1.3.2. Viral Structure and Function

Viruses are acellular entities composed of nucleic acid (DNA or RNA) enclosed in a protein coat known as a capsid. Some viruses possess an additional lipid envelope derived from the host cell membrane. The viral structure is extremely diverse, ranging from simple helical and icosahedral shapes to more complex structures (Maier et al., 2015).

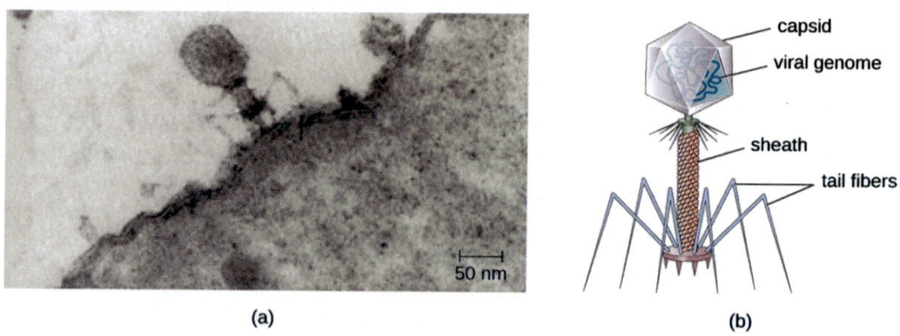

Figure 1.4. *Virus structure.*

Source: https://courses.lumenlearning.com/suny-microbiology/chapter/viruses/.

Viruses lack the machinery for independent metabolism and replication, so they rely entirely on host cells to reproduce. The viral life cycle consists of attachment, penetration, uncoating, replication, assembly, and release. The interaction between viruses and host cells can result in a variety of outcomes, including cell lysis and latent infections (Murray et al., 2015).

1.3.3. Fungal Structure and Function

The fungal cell wall, which is made up of chitin, glucans, and other polysaccharides, provides rigidity and protection. Fungi can be either unicellular yeasts or multicellular molds, with filamentous structures called hyphae that aggregate into mycelium (Murray et al., 2015).

Fungi reproduce sexually and asexually through the production of spores. Sexual reproduction involves the fusion of specialized cells to form zygospores, ascospores, or basidiospores, depending on the fungal class (Gest, 2004). Asexual reproduction involves the formation of conidia or sporangia.

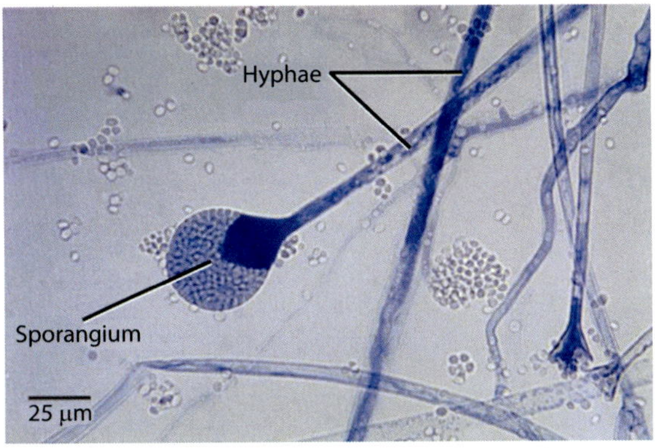

Figure 1.5. *Structure of fungi.*

Source: https://courses.lumenlearning.com/suny-osbiology2e/chapter/characteristics-of-fungi/).

Fungi are heterotrophic, obtaining nutrients via extracellular digestion. They produce enzymes that degrade complex organic materials into simpler compounds that can be absorbed. This saprophytic lifestyle is important for nutrient cycling and decomposition in ecosystems.

1.3.4. PROTOZOAL STRUCTURE AND FUNCTION

Protozoa cells have a nucleus, an endoplasmic reticulum, a Golgi apparatus, mitochondria, and, in some cases, specialized organelles such as contractile vacuoles for osmoregulation. Protozoa use a variety of nutritional strategies, including phagotrophy (particulate food ingestion), osmotrophy (dissolved nutrient uptake), and mixotrophy (a combination of both strategies). They play important roles in aquatic and soil ecosystems, contributing to the microbial loop by preying on bacteria and other microorganisms (Maier et al., 2015).

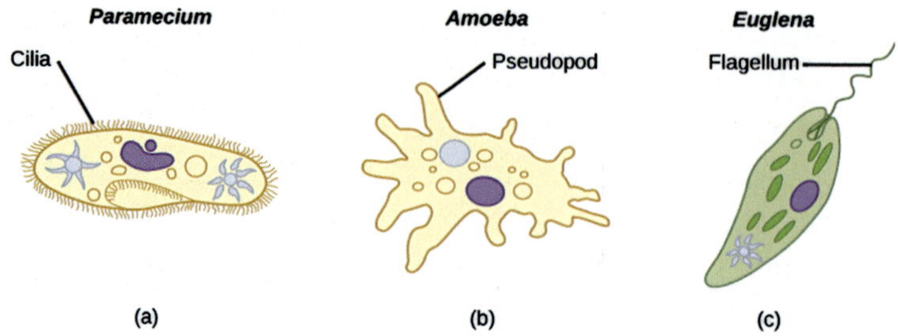

Figure 1.6. *Protozoa structure.*

Source: https://courses.lumenlearning.com/suny-wmopen-biology2/chapter/characteristics-of-protists/.

1.3.5. Algal Structure and Function

Algal cells have chloroplasts, which contain pigments such as chlorophyll that capture light energy. Algal cell walls are made up of cellulose, pectin, and other polysaccharides that vary by algal group. Algae reproduce in two ways: sexually and asexually. Asexual reproduction involves cell division, fragmentation, or spore formation, whereas sexual reproduction involves gamete fusion. Algae are primary producers in aquatic ecosystems, producing organic matter and oxygen via photosynthesis (Madigan et al., 2018).

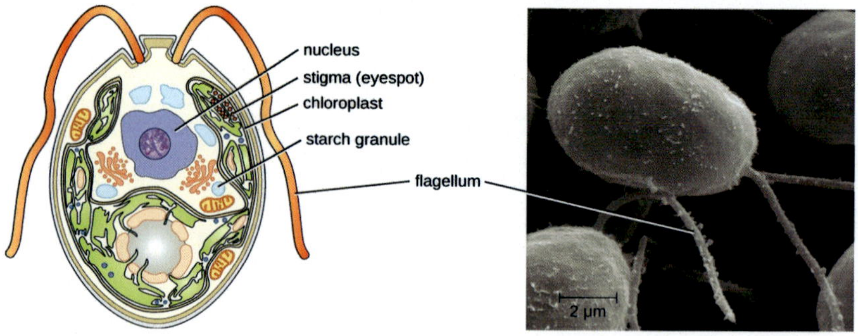

Figure 1.7. *Algae cell structure.*

Source: https://courses.lumenlearning.com/suny-microbiology/chapter/algae/.

1.3.6. Archaeal Structure and Function

In contrast to bacterial and eukaryotic membranes, archaeal cell membranes contain ether-linked lipids, which provide stability in extreme environments. Archaea's cell wall composition varies; some have pseudopeptidoglycan, while others do not have a cell wall at all (Maier et al., 2015). Archaeal flagella (archaella) are structurally distinct from bacterial flagella and play a role in motility. Archaea have distinct mechanisms for DNA replication, transcription, and translation, indicating their evolutionary divergence from bacteria (Madigan et al., 2018).

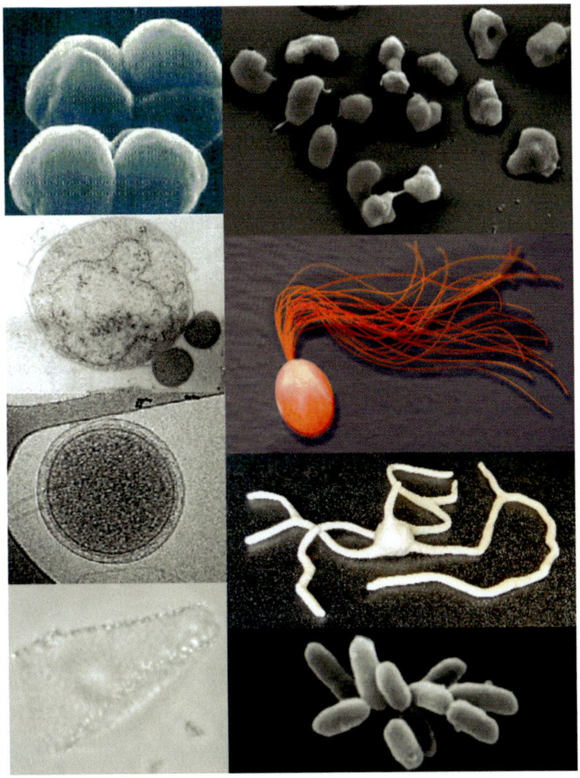

Figure 1.8. *The structures of archaeal cell.*

Source: By Maulucioni – Own work, CC BY-SA 4.0, https://commons.wikimedia.org/w/index.php?curid=89047952.

1.4. MICROBIAL GROWTH AND REPRODUCTION

Microbial growth and reproduction processes differ significantly among bacteria, archaea, fungi, protozoa, algae, and viruses, reflecting their diverse biological characteristics and ecological niches.

1.4.1. Bacterial Growth and Reproduction

Bacterial growth is typically achieved through binary fission, a simple asexual reproduction method. In binary fission, a single bacterial cell replicates its DNA, grows, and splits into two genetically identical daughter cells. This process has several key steps, including DNA replication, cell elongation, septum formation, and cytokinesis (Madigan et al., 2018).

The growth curve, which has four phases: lag, log (exponential), stationary, and death, is commonly used to describe the rate of bacterial growth. During the lag phase, bacteria adjust to their surroundings by synthesizing necessary enzymes and molecules. The log phase is distinguished by rapid and exponential growth. In the stationary phase, nutrient depletion and waste accumulation inhibit further growth, leading to a plateau. Finally, the death phase occurs as cells die at an exponential rate due to adverse conditions (Roller et al., 2016).

Environmental factors such as temperature, pH, and nutrient availability have a significant impact on bacterial growth. Extremophiles thrive in harsh environments like acidic hot springs or high-temperature hydrothermal vents. Bacteria can also form biofilms, which are complex communities that attach to surfaces and provide increased resistance to environmental stresses and antibiotics (Murray et al., 2015).

1.4.2. Viral Growth and Reproduction

The steps involved in viral growth and reproduction are attachment, penetration, uncoating, replication, assembly, and release. Viruses attach to host cells via specific receptors and penetrate their membranes. Following uncoating, the viral genome is released into the host cell, hijacking its machinery to replicate viral components. New virions are assembled and then released, frequently resulting in host cell lysis. This process ensures that the virus propagates within the host organism. Viral replication mechanisms are essential for comprehending viral pathogenesis and developing antiviral treatments (Murray et al., 2015).

1.4.3. Fungal Growth and Reproduction

Fungi reproduce both asexually and sexually, which helps them adapt and thrive in the environment. Asexual reproduction occurs via spore formation, budding, or fragmentation. Spore formation involves the production of numerous asexual spores, such as conidia or sporangiospores, which disperse and germinate into new mycelium (Roller et al., 2016). Budding is a common process in yeasts that involves the formation of a new cell from the parent cell. Fragmentation is the process of separating mycelium into pieces that can grow into new individuals. Fungi reproduce sexually by fusing haploid cells to form diploid zygospores, ascospores, or basidiospores, depending on their fungal group. This process increases genetic diversity, which improves the population's ability to adapt to changing environments (Maier et al., 2015; Murray et al., 2015).

1.4.4. Protozoal Growth and Reproduction

Protozoa reproduce both asexually and sexually, using binary fission, budding, and schizogony (multiple fission). Schizogony occurs when the nucleus divides multiple times before the cell splits into numerous daughter cells. Sexual reproduction in protozoa, such as conjugation in ciliates, is the exchange of genetic material between two cells, resulting in genetic recombination and diversity (Murray et al., 2015).

1.4.5. Algal Growth and Reproduction

Algae reproduce asexually and sexually. Asexual reproduction occurs through cell division, fragmentation, or spore formation. Binary fission is common in unicellular algae, whereas multicellular algae can reproduce through fragmentation or asexual spores. In algae, sexual reproduction occurs through the fusion of gametes, which results in genetic recombination and the formation of zygotes that develop into new individuals (Maier et al., 2009).

1.4.6. Archaeal Growth and Reproduction

Archaea, like bacteria, reproduce primarily through binary fission, although some species use budding or fragmentation. Temperature, salinity, and pH have similar effects on archaeal growth rates and environmental adaptations. Notably, many archaea are extremophiles, thriving in environments unsuitable for most other life forms. For example, halophilic archaea flourish in high-salinity environments, while thermophilic archaea thrive in extreme heat (Maier et al., 2009).

1.5. MICROBIAL METABOLISM

Microbial metabolism refers to the complex biochemical processes that microorganisms use to generate energy and construct cellular components. These processes are critical for microorganism survival, growth, and reproduction, and have an impact on their ecological roles and biotech applications.

1.5.1. Catabolism and Anabolism

Microbial metabolism is classified into two major categories: catabolism and anabolism. Catabolism is the breakdown of complex organic molecules into simpler ones, which releases energy that is used for cellular activities. Anabolism, on the other hand, uses this energy to create complex molecules from simpler ones, allowing cells to grow and survive (Madigan et al., 2018).

1.5.2. Energy Production

Microorganisms employ various metabolic pathways to produce energy. The most common are respiration and fermentation.

- **Respiration** can be either aerobic or anaerobic. Aerobic respiration uses oxygen as the terminal electron acceptor in the electron transport chain (ETC) to produce ATP via oxidative phosphorylation. This process combines glycolysis, the citric acid cycle, and the ETC to generate up to 38 ATP molecules per glucose molecule (Maier et al., 2009). Certain bacteria and archaea use anaerobic respiration, which involves terminal electron acceptors other than oxygen, such as nitrate, sulfate, or carbon dioxide. This process is less efficient than aerobic respiration, but it allows microorganisms to thrive in oxygen-deprived conditions (Madigan et al., 2018).
- **Fermentation** is an anaerobic process in which organic molecules, usually carbohydrates, are partially oxidized to generate ATP. This process produces less energy than respiration, generating only two ATP molecules per glucose molecule via substrate-level phosphorylation. Ethanol, lactic acid, and various gases are typical fermentation products, depending on the microorganism and substrate (Murray et al., 2015).

1.5.3. Biosynthetic Pathways

Anabolism refers to the biosynthesis of essential cellular components such as nucleic acids, proteins, lipids, and polysaccharides. These pathways necessitate both energy, often in the form of ATP, and reducing power, typically NADPH (Murray et al., 2015).

Nucleotide biosynthesis is essential for DNA and RNA synthesis. Purines and pyrimidines are produced via complex pathways that include precursors, such as amino acids, carbon dioxide, and ribose-5-phosphate. The regulation of these pathways maintains balanced nucleotide pools for nucleic acid synthesis and repair (Murray et al., 2015).

Amino acid biosynthesis differs between microorganisms, with many capable of producing all essential amino acids. These pathways frequently involve transamination reactions, in which amino groups are transferred to carbon skeletons derived from intermediates in central metabolic pathways (Madigan et al., 2018).

Lipid biosynthesis is essential for membrane formation and energy storage. Fatty acids are formed by the repeated addition of two-carbon units derived from acetyl-CoA. The resulting fatty acids can be modified to produce complex lipids such as phospholipids and glycolipids, which are required for cellular membranes (Maier et al., 2009).

Polysaccharide biosynthesis is the polymerization of sugar monomers into complex carbohydrates like peptidoglycan in bacterial cell walls and glycogen for energy storage. These processes are tightly controlled to meet the cell's structural and energy requirements (Madigan et al., 2018).

1.6. MICROBIAL GENETICS

Microbial genetics, a branch of genetics, investigates the mechanisms of heritable information in microorganisms such as bacteria, archaea, viruses, and some fungi and algae. The area has grown greatly since the early twentieth century, thanks to advances in molecular biology, genomics, and bioinformatics. This comprehensive overview delves into the structure, function, variation, and applications of microbial genetics.

1.6.1. Genetic Organization and Structure

Microbial genomes are normally arranged into chromosomes and plasmids, with bacteria frequently carrying a single, circular chromosome as well as multiple plasmids. Genomic sequencing methods have significantly influenced our understanding of microbial genome architecture. For example, Franco-Duarte et al. (2019) describe how sequencing technology has opened up new possibilities for studying the microbiome, demonstrating the tremendous genetic variety of bacteria. Genomic islands, including pathogenicity islands, are important components of microbial genomes that aid in horizontal gene transfer (HGT). These mobile genetic components let microbial pathogens evolve by integrating foreign DNA. Thus, the findings emphasize the dynamic nature of microbial genomes, which are altered by both vertical and horizontal gene exchanges.

The following are the microbial genetic makeup of micro-organisms:

i. **Microbial Genetics of Bacteria:** Bacteria have one circular chromosome and frequently carry plasmids, which are tiny, circular DNA molecules. These plasmids can carry genes for antibiotic resistance, virulence factors, and metabolic activity. Bacteria reproduce asexually via binary fission; however, they can exchange genetic material via HGT methods such as transformation (free DNA uptake), transduction (bacteriophage transfer), and conjugation (direct transfer through cell contact) (Madigan et al., 2018).

ii. **Microbial Genetics of Archaea:** Archaea, like bacteria, have a circular chromosome and can carry plasmids. Their genetic machinery, as well as transcription and translation processes, are more similar to those of eukaryotes than bacteria. HGT is common in archaea, which contributes to their adaptation in severe settings. Archaeal biology is poorly understood, but recent breakthroughs in sequencing technologies have given information on their distinct genetic and metabolic pathways (Murray et al., 2015).

iii. **Microbial Genetics of Viruses:** Viruses have extremely different genomes, which can be made of either DNA or RNA and can be single- or double-stranded. Viral genomes can be either linear or circular, and their size varies substantially. Viruses exploit host cells for reproduction, hijacking the host's machinery to duplicate their genetic material. Mutations, recombination, and reassortment contribute to the genetic diversity of viruses, particularly RNA viruses with high mutation rates (Madigan et al., 2018).

iv. **Microbial Genetics of Fungi:** Fungi contain complicated eukaryotic genomes made up of several linear chromosomes. They reproduce sexually and asexually, with sexual reproduction consisting of the union of haploid nuclei to generate diploid cells, followed by meiosis. Fungal genetics encompasses complex regulatory networks and gene expression processes. HGT is less widespread in fungi than in bacteria, but it is important for adapting to environmental challenges and acquiring new metabolic capacities (Maier et al., 2009).

v. **Microbial Genetics of algae:** Algae, a varied category of photosynthetic eukaryotes, contain complex genomes that include numerous linear chromosomes. Their genetic research is critical to understanding primary production and biogeochemical cycles. Algae reproduce both sexually and asexually, and their genetics include systems that regulate gene expression in response to environmental changes. Endosymbiotic processes have influenced the evolution and diversification of algae by facilitating HGT (Murray et al., 2015).

1.6.2. Genetic Variation and Transfer

Mutations, recombination, and HGT methods such as transformation, transduction, and conjugation all contribute to microbial genetic variety. Franco-Duarte et al. (2019) present a thorough examination of the molecular mechanisms underlying genetic variation and its evolutionary implications. He highlights the importance of chemical and physical mutagens in creating genetic variety, which is required for microbial adaptability and evolution. HGT is critical for propagating antibiotic resistance throughout microbial communities. Researchers have investigated the genetic and pharmacological mechanisms that confer antibiotic resistance, focusing on the involvement of plasmids and transposons in spreading resistance genes. This transfer of genetic material poses a significant problem in treating infectious diseases, needing continual research and surveillance (Franco-Duarte et al., 2019).

1.6.3. Microbial Genomics and Proteomics

Advances in sequencing technologies have transformed microbial genomics, allowing for thorough analysis of microbial genomes and metagenomes. Metagenomics, or the study of genetic material extracted directly from environmental samples, has revealed the genetic potential of uncultured microbes. Metagenomics applications in the identification of novel genes and metabolic pathways. This method has helped to identify novel antibiotics, enzymes, and other bioactive chemicals.

Proteomics, the study of all proteins expressed by a genome, supplements genomic data by providing information about protein function and interaction. Proteomic technologies, such as mass spectrometry (MS), have helped us comprehend microbial physiology and metabolism. Gest (2004) emphasizes the importance of genetics and proteomics in understanding microbial interactions during food fermentation. These technologies make fermentation processes more efficient, resulting in increased food and biofuel output.

1.6.4. Pathogenicity and Antibiotic Resistance

The genetic basis of microbial pathogenicity is the acquisition and production of virulence factors, which allow bacteria to infect hosts while evading immune responses. Numerous virulence genes have been found through genomic investigations, serving as targets for the development of novel treatments and vaccines. Pathogenicity islands play an extremely important role in the evolution of bacterial pathogens. These genomic areas include clusters of virulence genes that can be passed down between bacteria, increasing their pathogenic potential.

Antibiotic resistance is a major public health hazard caused by the genetic adaptability of microorganisms. For example, Gest (2004) investigated the ecological and molecular aspects of microbial interactions, focusing on how resistance genes spread in microbial communities. Understanding the genetic mechanisms of resistance aids in the development of strategies to limit the spread of resistant ailments.

1.6.5. Biotechnological Applications

Microbial genetics has several biotechnological applications, such as genetic engineering, industrial microbiology, and environmental microbiology. Microbial genomes can now be precisely modified thanks to genetic engineering tools like CRISPR-Cas systems. The application of genetically altered microbes in bioremediation demonstrates their ability to break down environmental contaminants. This application highlights the usefulness of microbial genetics in tackling environmental issues (Sieuwerts et al., 2008).

Industrial microbiology uses genetically modified bacteria to produce medications, biofuels, and other important items. The use of genomic and proteomic techniques optimizes microbial activities in food fermentations, increasing productivity and quality. These improvements have a tremendous economic and societal impact.

Environmental microbiology studies the role of microorganisms in biogeochemical cycles and ecosystem functioning. Metagenomic studies have helped us better understand microbial diversity and function in natural environments. This knowledge helps to guide conservation efforts and sustainable natural resource management (Franco-Duarte et al., 2019).

1.7. MICROBIAL ECOLOGY

Microbial ecology, a pivotal branch of microbiology, delves into the interactions, diversity, and functions of microorganisms within their environments. This field has garnered significant attention due to its implications for health, agriculture, biotechnology, and environmental sustainability. The following discussion, highlights microbial diversity and community structure, microbial habitats and niches, biogeochemical cycles, microbial interactions with plants and animals, and environmental and applied microbial ecology.

1.7.1. Community Structure

Microbial community patterns and diversity are essential to comprehending the stability and functionality of ecosystems in microbial ecology. These communities, which include fungi, viruses, bacteria, and archaea, display intricate spatial arrangements and interactions that are impacted by interspecies connections, resource availability, and environmental factors. Ecosystem resilience and productivity are supported by species richness and evenness indices, which measure diversity within these groups. Increased metabolic capacity and ecological flexibility are frequently correlated with high microbial diversity. These traits are essential for the cycling of nutrients, the breakdown of organic matter, and the reduction of environmental stress. Modern molecular methods, like high-throughput sequencing and metagenomics, have completely changed how these complex and dynamic microbial assemblages are studied (Lederberg and Relman, 2009).

1.7.2. Microbial Habitats and Niches

Microorganisms live in a wide variety of settings, including terrestrial and aquatic ecosystems, each with its own advantages and drawbacks. Microbial life is abundant in aquatic habitats, such as lakes, rivers, and seas, and it is essential to primary production and nutrient cycling (Simon et al., 2002). Soil microbiota has an important role in the breakdown of organic matter, cycling of nutrients, and plant health in terrestrial environments. Microorganisms known as extremophiles, which can survive in harsh environments like high salinity, high temperature, or high pressure, offer valuable insights into the boundaries of life and may find use in biotechnology (Madigan et al., 2018).

1.7.3. Biogeochemical Cycles

Microorganisms are integral to biogeochemical cycles, driving the transformation and movement of elements such as carbon, nitrogen, sulfur, and phosphorus. In the carbon cycle, microbes mediate processes like carbon fixation, decomposition, and methanogenesis, influencing atmospheric CO_2 levels and climate change. The nitrogen cycle involves microbial processes such as nitrogen fixation, nitrification, and denitrification, which are critical for maintaining soil fertility and ecosystem productivity (Madigan et al., 2018). Similarly, microbial activity regulates sulfur and phosphorus cycles, impacting soil and water quality.

1.7.4. Microbial Interactions with Plants and Animals

Microbial interactions with plants and animals are essential for ecosystem health and function. Microbial communities in the rhizosphere, or zone of soil impacted by plant roots, improve nutrient intake, boost plant growth, and defend against diseases. Symbiotic connections, such as those between legumes and nitrogen-fixing bacteria, are critical to sustainable agriculture. The gut microbiome affects health, digestion, and disease resistance in animals, especially humans. Microbial community disruptions can cause obesity and inflammatory disorders (Maier et al., 2015; Murray et al., 2015).

1.7.5. Environmental and Applied Microbial Ecology

The study of environmental microbial ecology is concerned with the functions of microbes in both natural and artificial environments, with a particular emphasis on agriculture, bioenergy, and bioremediation. By using microorganisms' metabolic potential to break down environmental contaminants, bioremediation provides damaged sites with long-lasting solutions. Beneficial bacteria are used in agriculture as bio-pesticides and bio-fertilizers to increase crop yields while lowering chemical inputs. Since microbial contamination poses serious dangers to public health, understanding the ecology of microbes in drinking water systems is essential to maintaining the safety and purity of the water (Lederberg and Relman, 2009).

1.8. MICROBIAL DIVERSITY

Microbial diversity refers to a large array of microorganisms. This section highlights a range of studies with significant national implications. Microorganisms such as fungi, bacteria, and viruses exhibit a wide range of natural variability. It is a natural resource that is not visible. Although microorganisms are numerous in the cosmos, their exact profile is often unclear (Timmis, 2002). Their tiny nature could be the explanation. For this reason, they remained untouched until roughly 300 years ago. Microorganisms are responsible for the vast diversity of life on Earth. Animals outnumber plants, vertebrates, and insects combined. Microorganisms have been around for many decades. They can react with a wide range of substances, including metals, acids, petroleum, and natural gas, all of which are harmful to humans.

1.8.1. Taxonomic Diversity

Taxonomic diversity in microbiology refers to the division of bacteria into various groups based on genetic and phenotypic traits. Bacteria are classified into various phyla based on their simple cell structure, including Proteobacteria, Firmicutes, and Actinobacteria. Archaea, which thrives in harsh settings, includes groups such as *Euryarchaeota* and *Crenarchaeota*. Fungi, which include yeasts and molds, are vital for nutrient cycling and decomposition. Single-celled eukaryotes known as protozoa play an important role in aquatic ecosystems. Algae, both macro and micro, contribute greatly to photosynthesis in the marine and freshwater ecosystems. Viruses, despite not being cellular, profoundly influence microbial population dynamics and genetic exchange. The study of microbial taxonomic diversity is crucial for understanding ecological interactions, evolutionary relationships, and biotechnological applications (Madigan et al., 2018).

1.8.2. Genetic Diversity

Genetic diversity in microorganisms is an important component of microbiology, representing the wide range of genetic variants found within microbial populations. This diversity is critical for microbes' ability to adapt and survive in a variety of situations. Mutations, HGT, and recombination all contribute to genetic variety, allowing microbes to develop quickly and adapt to changing environments. This diversity enhances their roles in ecological processes like as nutrient cycling, symbiosis, and disease. Studies have indicated that genetic variety improves microbial community resilience and functionality, providing a buffer against environmental perturbations. Additionally, it plays a significant role in biotechnology and medicine, as diverse microbial genes and metabolic pathways are exploited for developing novel antibiotics, enzymes, and industrial processes (Timmis, 2002). Understanding and preserving microbial genetic diversity is therefore essential for both ecological sustainability and biotechnological advancement.

1.8.3. Functional Diversity

Microorganisms have remarkable functional diversity and serve vital functions in a variety of ecosystems. Their metabolic adaptability allows them to live in a wide range of settings, including extreme thermophilic environments and deep-sea hydrothermal vents. Microorganisms play an important role in nutrient cycling by performing nitrogen fixation, carbon sequestration, and sulfur oxidation, all of which are required to maintain ecological equilibrium. Their capacity to break down organic contaminants makes them extremely useful in bioremediation techniques. Microorganisms also have an impact on human health, acting as commensal flora that promotes digestion and immunity, as well as pathogens that cause disease. Microorganisms' functional variety is based on their genetic flexibility, which enables fast evolution and acquisition of novel metabolic pathways. This diversity not only underscores the ecological significance of microorganisms but also their potential in biotechnology and medical applications, offering innovative solutions for environmental and health challenges (Maier et al., 2015; Murray et al., 2015).

1.8.4. Ecological Diversity

Microbiology's ecological diversity is immense, including a wide range of habitats and metabolic capabilities. Microorganisms such as bacteria, archaea, fungi, and protists live in a wide range of environments, from severe circumstances like hydrothermal vents and acidic hot springs to more typical habitats like soil, water, and humans. This diversity is critical to ecological balance and biogeochemical cycles. For example, bacteria play an important part in nitrogen fixation, which is required for plant growth. Archaea, with their distinct metabolic pathways, contribute to the carbon and sulfur cycles in harsh settings. Fungi break down organic substances and recycle nutrients back into ecosystems. Protists, as primary producers and predators, influence the aquatic food chain. The study of microbial diversity not only enhances our understanding of ecosystem functioning but also has implications for biotechnology and medicine (Maier et al., 2015).

1.8.5. Importance of Microbial Diversity

Microbiology's foundational concept, microbial variety has significant effects on human health, biotechnology, and ecological equilibrium. Numerous microorganisms, such as bacteria, fungi, viruses, and archaea, are essential to the decomposition of organic matter, cycling of nutrients, and symbiotic connections with plants and animals.

Diverse microbial communities improve ecosystem resilience by facilitating stressor and environmental change adaption. Using this diversity to biotechnology's advantage results in advances in bioengineering, agriculture, and medicine. For instance, novel antibiotics and enzymes are frequently derived from unique microbial strains. Furthermore, research on the human microbiome shows that microbial variety is essential for immune response regulation, illness prevention, and health maintenance. Thus, the study of microbial variety promotes breakthroughs in many scientific and industrial domains while also deepening our understanding of microbial ecology (Timmis, 2002).

1.9. MICROORGANISMS IN INDUSTRY

Microorganisms have been instrumental in the development of biotechnology, food production, environmental management, and pharmaceuticals, among other industrial applications. The industrial use of microorganisms takes advantage of their varied metabolic capacities to generate a variety of goods, including enzymes, biofuels, biofertilizers, and medications (Olempska-Beer et al., 2006). An outline of the major developments and uses of microorganisms in industry is highlighted in this section.

1.9.1. Food and Beverage Industry

The food and beverage sector uses a lot of microorganisms, mostly for fermentation operations. Fermentation is a process used to make yogurt, cheese, bread, beer, and wine. It is fueled by bacteria, yeasts, and molds. For instance, the baking and brewing industries depend heavily on *Saccharomyces cerevisiae*, also referred to as baker's yeast. In a similar vein, dairy products like yogurt and cheese cannot be produced without lactic acid bacteria (LAB) like *Lactobacillus* and *Bifidobacterium* (Olempska-Beer et al., 2006).

Enzyme manufacturing for food processing has been considerably improved by the introduction of recombinant DNA technology. Enzymes including lipases, proteases, and amylases are produced more effectively by recombinant microorganisms and are crucial for a variety of food processing applications.

1.9.2. Pharmaceutical Industry

Microorganisms play an important role in the pharmaceutical business, producing antibiotics, vaccines, and other medicinal substances. Streptomyces, a genus of actinobacteria, is well-known for producing more than two-thirds of clinically relevant natural antibiotics, such as streptomycin and erythromycin. Advances in genetics and metabolic engineering have allowed these microbes to be optimized to increase yield while lowering production costs (Franco-Duarte et al., 2019).

1.9.3. Industrial Enzymes

Industrial enzyme manufacturing via microbial fermentation has transformed industries such as textiles, leather, detergents, and paper. Microbial enzymes have several advantages, including high specificity, activity under mild environments, and ease of genetic manipulation. For example, microorganisms such as Trichoderma and Bacillus species produce enzymes such as cellulases and amylases, which are used in the bioconversion of biomass to biofuels and other value-added products (Ōmura et al., 2001).

1.9.4. Agricultural Applications

Microorganisms are used in agriculture as biofertilizers and insecticides, encouraging environmentally friendly farming techniques. Biofertilizers, including microorganisms such as Rhizobium and Azotobacter, improve soil fertility by fixing atmospheric nitrogen. Mycorrhizal fungi develop symbiotic associations with plant roots, which enhances nutrient uptake and growth. These microbial inoculants are increasingly being utilized to lessen reliance on chemical fertilizers and pesticides, hence decreasing environmental effects (Murray et al., 2015).

1.9.5. Environmental Biotechnology

Microorganisms are also used in environmental biotechnology for bioremediation and waste management. Halophilic and halotolerant bacteria, which flourish in high-salt conditions, are used in the bioremediation of saline and hypersaline environments contaminated by organic contaminants. In addition, microbes are used to remediate industrial effluent and detoxify dangerous compounds (Ōmura et al., 2001).

1.9.6. Biofuel Production

Biofuels, such as biodiesel and bioethanol, are gaining popularity as a sustainable alternative to fossil fuels. Oleaginous bacteria, which can collect lipids, are being investigated for biodiesel production. These microorganisms, including specific species of algae and yeasts, are bred to improve lipid accumulation and conversion efficiency (Franco-Duarte et al., 2019).

1.9.7. Future Prospects and Challenges

Synthetic biology and systems biology techniques must be integrated if microbial applications are to continue in the industrial sector. These cutting-edge techniques improve the synthesis of targeted chemicals by enabling the design and development of novel metabolic pathways. To fully utilize the potential of microorganisms in the industry, however, challenges like public perception, scalability, and regulatory concerns must be managed (Ōmura et al., 2001).

1.10. Microorganisms in Medicine

Microorganisms, which include bacteria, viruses, fungi, and protozoa, have become more and more important in the medical sector, especially since the year 2000. The understanding of how microbes affect human health has advanced significantly throughout this time, resulting in ground-breaking discoveries in illness diagnosis, treatment, and prevention.

1.10.1. The Role of Microorganisms in Disease Pathogenesis

Microorganisms have a crucial role in the etiology of many diseases. Bacterial pathogens like Staphylococcus aureus and *Escherichia coli* are well-known for producing infections ranging from mild skin problems to potentially fatal systemic illnesses. Advances in genomic sequencing have revealed the genetic mechanisms that contribute to their virulence and antibiotic resistance, a subject thoroughly covered by Davies and Davies (2010). Their research highlights the adaptive methods that these bacteria use, such as HGT, which hampers treatment attempts and needs the development of new antimicrobial treatments.

1.10.2. Microbiome and Human Health

The billions of bacteria that live mostly in the gut that make up the human microbiome have become significant markers of both health and disease. Since its launch in 2007, the Human Microbiome Project has offered an abundance of information about the make-up and purposes of the microbial communities that inhabit the human body. Numerous physiological functions, such as digestion, immunological response, and even mental health, are influenced by the microbiome. Obesity, depression, and inflammatory bowel disease have all been connected to dysbiosis, or the unbalance of microbial communities(Davies and Davies, 2010).

1.10.3. Microorganisms in Diagnostics

The development of molecular diagnostic tools has transformed pathogen detection and identification. Polymerase chain reaction (PCR) and next-generation sequencing (NGS) technologies enable the rapid and accurate detection of infections, which is crucial for effective therapy. For example, Sanders et al. (2013) proved the effectiveness of NGS in finding novel viruses, such as the Lujo virus, which causes hemorrhagic fever. These improvements have considerably improved infectious illness treatment by allowing for quick and exact diagnosis, which facilitates appropriate therapeutic measures.

1.10.4. Antibiotic Resistance

Antibiotic resistance is one of today's most significant medical challenges. The abuse and overuse of antibiotics have resulted in the rise of multidrug-resistant organisms (MDROs), which pose a substantial danger to world health. The World Health Organization's 2014 report on antibiotic resistance emphasizes the pervasive scope of the problem and asks for coordinated global action. Antibiotic resistance can be combated by a variety of measures, including the creation of new antibiotics, antibiotic stewardship programs, and alternative therapies like bacteriophage therapy (Davies and Davies, 2010).

1.10.5. Therapeutic Applications of Microorganisms

Regardless of their involvement in disease, microbes have interesting therapeutic uses. Probiotics, or helpful bacteria, such as *Lactobacillus* and *Bifidobacterium*, have been extensively researched for their health advantages. Sanders et al. (2013) found that probiotics can aid with irritable bowel syndrome (IBS) and antibiotic-associated diarrhea. Furthermore, oncolytic viruses, which are designed to selectively infect and kill cancer cells, represent a novel approach to cancer therapy. Recent clinical trials have produced positive results in the use of oncolytic viruses to treat a variety of malignancies, underlining its potential as a supplemental therapy to established treatments (Sanders et al., 2013).

1.10.6. Vaccines and Immunization

Vaccination is one of the most effective ways to avoid infectious diseases. The creation of novel vaccinations and the modification of existing ones have proven critical in controlling diseases such as influenza, human papillomavirus (HPV), and hepatitis B. The launch of mRNA vaccines, such as the COVID-19 vaccines created by Pfizer-BioNTech and Moderna, represents a significant breakthrough. Polack et al. (2020) found that mRNA vaccines were highly effective and safe in preventing COVID-19, demonstrating the technology's potential to rapidly respond to new infectious illnesses.

Microorganisms' future in medicine looks promising. The combination of microbiome research with personalized medicine shows the potential for generating tailored treatments based on an individual's microbial composition. Furthermore, synthetic biology has the potential to create microbes for therapeutic applications, such as providing critical nutrients or breaking down harmful chemicals in the body. The ethical and safety aspects of these technologies will be critical to their effective application (Davies and Davies, 2010).

CHAPTER 2

MICROBIAL TECHNIQUES

Contents

Microbial methods are critical procedures in microbiology that allow for the isolation, identification, and characterization of microorganisms. These strategies cover a wide range of disciplines, each of which is crucial to improving our understanding of microbial life and its applications in fields as diverse as medicine, agriculture, and environmental science Microbial culturing, microscopy, and molecular approaches are some of the most fundamental techniques (Murray et al., 2015).

Microbial culturing is the process of growing microorganisms in controlled circumstances using various media types, allowing researchers to study microbial physiology and identify specific species. Pure cultures are frequently isolated from mixed microbial populations using techniques like streak plating, pour plating, and spread plating (Murray et al., 2015).

Microscopy, which includes light microscopy, fluorescence microscopy, and electron microscopy, gives critical information about microbial morphology and structure. Microscopy advances, such as confocal laser scanning microscopy, have considerably improved our ability to see microorganisms in their natural environment (Madigan et al., 2018).

Molecular techniques like polymerase chain reaction (PCR), genomic sequencing, and metagenomics have transformed microbiology by allowing for extensive investigation of microbial genomes and the study of microbial communities without the requirement for culturing. These techniques enable the genetic identification of microorganisms, aiding the discovery of new species and a better knowledge of microbial diversity and evolution (Murray et al., 2015).

These approaches constitute the foundation of current microbiological research, pushing advancements in clinical diagnostics, biotechnology, and ecological studies.

2.1. MICROSCOPIC EXAMINATION OF MICROORGANISMS

Microscopic examination allows for direct observation of microorganisms' morphology, structure, and behavior. This practice is critical for identifying and comprehending the enormous diversity of microbial life, which includes bacteria, archaea, fungi, protozoa, and viruses. To improve our visualization and understanding of microorganisms, various microscopy techniques have been developed, each with its own set of capabilities and applications (Beveridge et al., 2007).

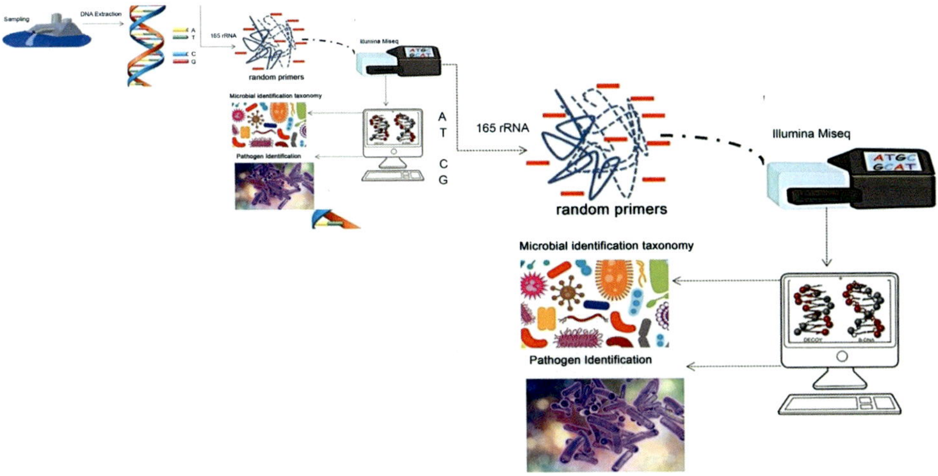

Figure 2.1. *General workflow of bacterial profiling.*

Source: Oluseyi Osunmakinde C, Selvarajan R, Mamba BB, Msagati TAM. (2019). Profiling bacterial diversity and potential pathogens in wastewater treatment plants using high-throughput sequencing analysis. Microorganisms, 7(11):506. https://doi.org/10.3390/microorganisms7110506.

2.1.1. Light Microscopy

Light microscopy is the most commonly used technique in microbiology. It uses visible light to illuminate samples, which are then magnified using a series of lenses. Standard light microscopes can magnify specimens by up to 1,000 times, allowing for the study of cellular structures and arrangements. Bright-field, dark-field, phase-contrast, and differential interference contrast (DIC) microscopy techniques all provide different ways to visualize microorganisms. Bright-field microscopy, the most basic type, frequently requires staining techniques to improve contrast and make structures such as cell walls, nuclei, and flagella more visible (Da Silva et al., 2018).

2.1.2. Electron Microscopy

Electron microscopy, which includes transmission electron microscopy (TEM) and scanning electron microscopy (SEM), achieves much higher resolution than light microscopy by employing electron beams rather than light. TEM creates detailed images of microorganisms' internal structures by transmitting electrons through thin sections of samples. This allows for the observation of subcellular components like organelles, ribosomes, and viral particles at magnifications of up to 2 million times (Beveridge et al., 2007). In contrast, SEM uses electron beams to scan the surface of specimens, producing three-dimensional images that reveal surface details and textures.

2.1.3. Fluorescence Microscopy

Fluorescence microscopy makes use of fluorophores, which are fluorescent molecules that emit light when excited at specific wavelengths. This technique significantly improves microorganism visualization, allowing for the precise and sensitive detection of specific cellular components and processes. Fluorescent dyes and genetically encoded fluorescent proteins, such as green fluorescent protein (GFP), can label a variety of cellular structures, making it easier to study living cells and dynamic processes (Murray et al., 2015). Confocal laser scanning microscopy, which focuses on thin sections of specimens, improves resolution and produces three-dimensional images.

2.1.4. Advanced Techniques

Advances in microscopy have resulted in the development of techniques that go beyond traditional limitations. Super-resolution microscopy, which includes methods such as stimulated emission depletion (STED) and photoactivated localization microscopy (PALM), overcomes light microscopy's diffraction limit and provides nanometer-scale resolution. These methods are invaluable for studying the ultrastructure of microbial cells and molecular interactions in unprecedented depth (Murray et al., 2015; Da Silva et al., 2018).

2.1.5. Staining Methods

Staining techniques are essential for microscopic examination because they increase contrast and enable the visualization of specific microorganism components. Hans Christian Gram developed the Gram stain, which uses cell wall properties to distinguish between Gram-positive and Gram-negative bacteria. Other stains, such as acid-fast, endospore, and capsule stains, target specific cellular structures and help identify and classify microorganisms (Beveridge et al., 2007; Da Silva et al., 2018).

2.2. CULTURING TECHNIQUE

Microbiology relies heavily on culturing techniques to isolate, identify, and study microorganisms. These methods rely on nutrient-rich media to provide optimal conditions for microbial growth.

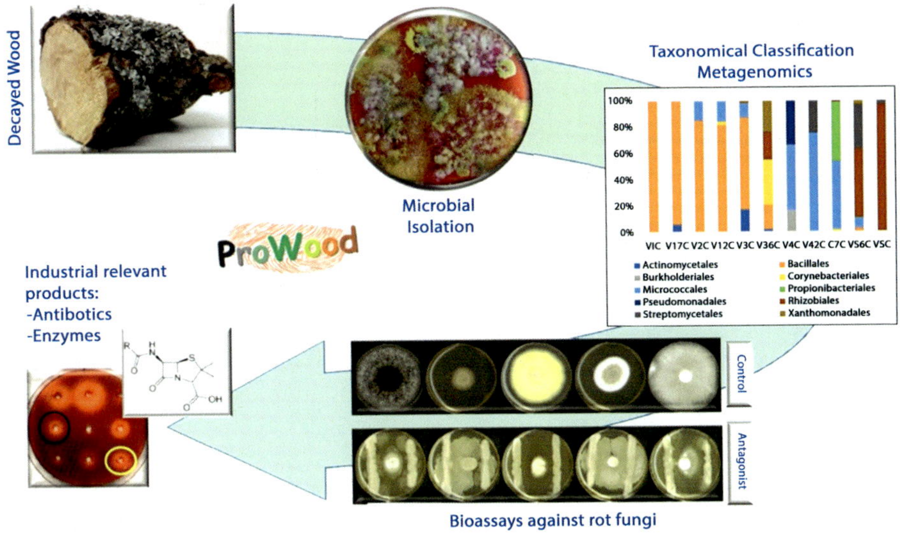

Figure 2.2. *Scheme of the optimized protocol for the cultivation of previously uncultured microorganisms from forest soil.*

Source: Velasco-Rodríguez Ó, Fil M, Heggeset TMB, Degnes KF, Becerro-Recio D, Kolsaková K, Haugen T, Jønsson M, Toral-Martínez M, García-Estrada C, et al. (2022). Characterization of microbial diversity in decayed wood from a spanish forest: an environmental source of industrially relevant microorganisms. Microorganisms, 10(6):1249. https://doi.org/10.3390/microorganisms10061249.

The following are the most commonly used culturing techniques in microbiology:

2.2.1. Streak Plate Method

The streak plate method is a key microbiological technique for isolating pure bacterial colonies from mixed cultures. Using a sterile loop, spread a small inoculum across the surface of an agar plate in a series of streaks. Each successive streak dilutes the sample, allowing individual cells to form separate colonies. This method is critical for obtaining isolated colonies for later identification and analysis of microbial species (Atlas, 2010).

2.2.2. Spread Plate Method

The spread plate method is a fundamental technique for isolating and quantifying microorganisms. This procedure involves using a sterile spreader to evenly distribute a diluted microbial sample across the surface of an agar plate. The technique is especially useful for counting colony-forming units (CFUs) from serial dilutions, which aids in the determination of microbial concentrations in samples. It ensures discrete colony growth, which is critical for future microbial identification and analysis (Beveridge et al., 2007).

2.2.3. Selective and Differential Media

Selective media contain agents that prevent the growth of some microorganisms while allowing others to grow. Differential media contain indicators that reveal specific metabolic activities or properties of microorganisms. These media types are critical for isolating and identifying specific microbial species based on physiological characteristics. Examples include MacConkey agar for distinguishing lactose fermenters from non-fermenters and Mannitol salt agar for isolating Staphylococcus species (Atlas, 2010).

2.2.4. Enrichment Cultures

Enrichment cultures employ selective media and growth conditions to promote the growth of specific microorganisms in a mixed population. This technique is especially useful for isolating fastidious organisms or those found in small numbers. Enrichment entails providing nutrients or environmental conditions that favor the target microorganism while inhibiting the growth of competitors (Madigan et al., 2018).

2.2.5. Pour Plate

In this method, a diluted microbial sample is combined with molten agar and poured into a petri dish. As the agar solidifies, microorganisms become immobilized within it, allowing colonies to grow throughout its depth. Because of its ability to grow under varying oxygen conditions, this technique is especially useful for counting viable microorganisms and isolating anaerobic bacteria (Atlas, 2010; Madigan et al., 2018).

2.2.6. Anaerobic Cultures

Anaerobic cultures are essential for growing microorganisms that thrive in oxygen-free environments. To eliminate oxygen, these techniques use specialized equipment such as anaerobic chambers or jars, as well as media containing reducing agents. Anaerobic culture is essential for studying obligate anaerobes, which play important roles in a variety of ecological and clinical settings, including gastrointestinal microbiota and pathogen research. Proper anaerobic conditions allow for the growth and accurate study of these microbes' metabolic and physiological properties (Atlas, 2010; Madigan et al., 2018).

2.2.7. Broth Cultures

Broth cultures are a technique for growing microorganisms in liquid nutrient media. This method enables the cultivation of large numbers of microbial cells and is critical for researching microbial growth kinetics and physiology. Broth cultures are kept under controlled conditions in flasks or tubes, allowing microorganisms to be propagated for further analysis such as biochemical assays and antimicrobial susceptibility testing. They are also essential for maintaining stock cultures and producing biomass for various experimental purposes (Atlas, 2010).

2.3. ISOLATION AND IDENTIFICATION OF MICROORGANISMS

The isolation and identification of microorganisms is critical for understanding microbial diversity, pathogenicity, ecology, and biotechnology applications. These processes employ a variety of sophisticated techniques for obtaining pure cultures and accurately identifying microbial species.

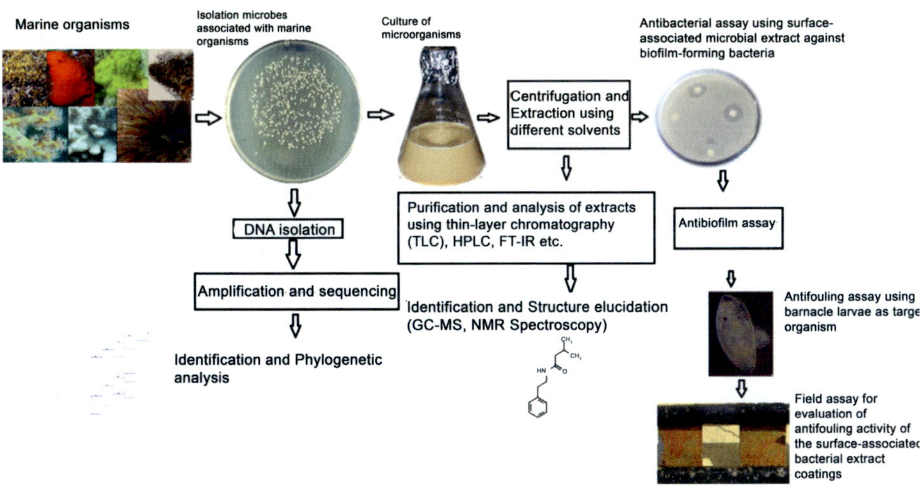

Figure 2.3. *Beach sand sample collection, processing, identification, and effect on HaCaT cells.*

Source: Subhan F, Shahzad R, Tauseef I, Haleem KS, Rehman AU, Mahmood S, Lee IJ. (2018). Isolation, identification, and pathological effects of beach sand bacterial extract on human skin keratinocytes in vitro. PeerJ., 6:e4245. doi: 10.7717/peerj.4245. PMID: 29441229; PMCID: PMC5807979.

We will review some of the culture techniques discussed in Section 2.2 above because they are also common isolation methods in microbiology.

- **Streak Plate Method:** The streak plate method uses a sterile inoculating loop to spread a small amount of microbial sample across the surface of an agar plate in a series of streaks. Each successive streak dilutes the sample more, allowing individual cells to be spatially separated and form discrete colonies during incubation. This method is required to obtain pure cultures for further analysis and characterization (Atlas, 2010).
- **Enrichment Cultures:** Enrichment cultures are essential in microbiology for isolating specific microorganisms from complex communities. By employing selective media and tailored growth conditions, researchers can enhance the proliferation of target organisms, particularly those that are difficult to cultivate or exist in minimal quantities. This technique strategically supplies nutrients and environmental factors that favor the desired microorganisms, effectively suppressing the growth of competing species. As a result, enrichment cultures serve as a powerful tool for studying and understanding the dynamics of microbial populations.
- **Pour Plate Method:** The pour plate method involves combining a diluted microbial sample with molten agar and pouring the mixture into a petri dish. As the agar solidifies, microorganisms become immobilized, allowing colonies to form throughout the medium. This method is especially useful for isolating and counting viable microorganisms, including anaerobes, because it can grow under a variety of oxygen conditions (Madigan et al., 2018).
- **Spread Plate Method:** Using a sterile spreader, a diluted microbial sample is evenly distributed across the surface of an agar plate in the spread plate method. This method allows for the counting of CFUs from serial dilutions, which makes it especially helpful for quantifying microorganisms. As per Beveridge et al. (2007), this approach guarantees distinct colony growth, an essential requirement for precise microbial identification.
- **Selective and Differential Media:** Selective media contain agents that prevent the growth of some microorganisms while allowing others to thrive. Differential media contain indicators that reveal distinct metabolic activities or properties, facilitating the isolation and identification of specific microbial species (Atlas, 2010).

Following the isolation of microorganisms and successful culturing, the process of identification begins based on several parameters of colonies. Common identification methods include:

- **Microscopic Examination:** Microscopy remains an important tool for identifying microorganisms based on their morphological characteristics. Light microscopy, fluorescence microscopy, and electron microscopy enable detailed visualization of microbial structures. Staining methods, such as Gram staining and acid-fast staining, provide important information about cell wall properties and aid in the initial classification of microorganisms (Beveridge et al., 2007).
- **Biochemical Tests:** Biochemical tests identify microorganisms by examining their metabolic and enzymatic properties. These tests evaluate the organism's ability to metabolize specific substrates, produce specific enzymes, and generate metabolic byproducts. For example, the catalase test distinguishes between catalase-positive and catalase-negative bacteria, whereas the oxidase test identifies organisms that produce cytochrome oxidase. Carbohydrate fermentation tests and the urease test help to classify microorganisms based on their metabolic profiles. These microbial physiology-based assays are critical for clinical diagnostics and microbial taxonomy, allowing for precise differentiation between closely related species (Madigan et al., 2018).
- **Molecular Techniques:** Molecular methods like PCR, genomic sequencing, and metagenomics have transformed microbial identification. PCR amplifies specific DNA sequences, enabling highly specific and sensitive detection and identification of microorganisms. Genomic sequencing provides comprehensive insights into microbial genomes, allowing for species identification and genetic understanding. Metagenomics, the study of genetic material extracted directly from environmental samples, enables the analysis of microbial communities without the need for culturing (Murray et al., 2015).
- **Mass Spectrometry:** Mass spectrometry (MS) identifies microorganisms by examining the distinct protein or lipid profiles of their cellular components. Microbiology frequently employs techniques such as Matrix-Assisted Laser Desorption/Ionization-Time of Flight (MALDI-TOF) MS. This method generates mass

spectra quickly and compares them to large databases, allowing for precise microbial identifications. MS is valued for its accuracy, speed, and ability to distinguish between closely related species (Murray et al., 2015).

- **Serological Methods:** Serological methods detect specific antigens or antibodies in patient samples. Common techniques include enzyme-linked immunosorbent assays (ELISA), agglutination tests, and immunofluorescence. These assays use the specificity of antibody-antigen interactions to identify pathogens based on their surface antigens or the host's immune response. Serological methods are especially useful for diagnosing viral, bacterial, and fungal infections because they provide quick and specific identification even when the microorganism is difficult to culture (Atlas, 2010; Madigan et al., 2018).

2.4. MICROBIAL STAINING TECHNIQUES

Microbial staining techniques are essential for visualizing and distinguishing microorganisms based on their morphological and structural properties.

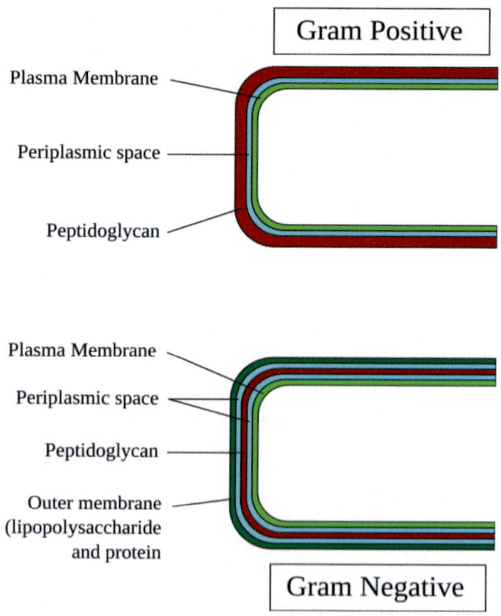

Figure 2.4. *Gram staining technique distinguishes based on the cell wall structure.*

Source: By Graevemoore at English Wikipedia, CC BY-SA 3.0, https://commons.wikimedia.org/w/index.php?curid=8477969.

Key staining techniques include:

2.4.1. Gram Staining

Gram staining, invented by Hans Christian Gram in 1884, is a key technique in microbiology for classifying bacteria based on their cell wall composition. The procedure consists of four main steps: crystal violet application, iodine treatment, alcohol decolorization, and safranin counterstaining. Gram-positive bacteria, with thick peptidoglycan layers, retain the crystal violet-iodine complex and appear purple, whereas Gram-negative bacteria, with thinner peptidoglycan and an outer membrane, lose the initial stain and appear pink after safranin treatment. This differential staining is essential for bacterial identification, as it guides diagnostic and treatment strategies (Beveridge, 2007).

2.4.2. Negative Staining

Negative staining is a microbiology technique that improves the visibility of delicate structures, such as bacterial capsules, by staining the background rather than the organism. This method uses an acidic dye such as India ink or nigrosin, which does not penetrate the cell. This creates a dark background against which the transparent, unstained cells and structures stand out as bright, contrasted silhouettes. Negative staining is especially useful for observing capsules, which are difficult to detect using positive staining techniques. This technique is critical for studying morphology and diagnosing encapsulated pathogens (Baltz et al., 2010).

2.4.3. Capsule Staining

Capsule staining is a specialized microbiology technique for visualizing bacterial capsules, which are polysaccharide layers that surround specific bacterial cells. Due to their non-ionic nature, capsules are difficult to stain using conventional methods. The technique typically involves combining negative staining, which stains the background, with a counterstain, such as crystal violet or safranin, to color the cells. The capsule appears as a clear halo surrounding the stained cell. This method is critical for identifying pathogenic bacteria because capsules increase virulence by protecting bacteria from phagocytosis and facilitating immune evasion (Beveridge et al., 2007).

2.4.4. Endospore Staining

Endospore staining is a microbiological technique for visualizing bacterial endospores, which are highly resistant, dormant structures produced by certain genera such as Bacillus and Clostridium. The most common method is the Schaeffer-Fulton technique, which involves staining with malachite green and then applying heat to penetrate the tough spore coat. The sample is then counterstained with safranin to highlight the vegetative cells. Endospores retain the green stain and appear as green ovals inside red or pink cells. This technique is critical for identifying spore-forming bacteria and understanding their role in survival in harsh environments (Beveridge et al., 2007).

2.4.5. Acid-Fast Staining

Acid-fast staining, also known as the Ziehl-Neelsen method, is an important technique in microbiology for identifying acid-fast bacteria, particularly those from the genus Mycobacterium. These bacteria have waxy, lipid-rich cell walls that resist acid-alcohol decolorization following carbol fuchsin staining. The procedure consists of applying carbol fuchsin with heat, followed by acid-alcohol decolorization and a counterstain, usually methylene blue. Acid-fast bacteria retain the red color of carbol fuchsin, resulting in bright red rods, whereas non-acid-fast cells absorb the blue counterstain. This method is essential for detecting tuberculosis and other mycobacterial infections (Atlas, 2010; Murray et al., 2015).

2.5. MICROBIAL BIOCHEMICAL TESTS

Microbial biochemical tests identify and differentiate microorganisms based on their biochemical activities. These tests exploit the metabolic and enzymatic capabilities of microorganisms, providing critical data for taxonomy, diagnostics, and research. The primary biochemical tests include:

2.5.1. Carbohydrate Fermentation Tests

Carbohydrate fermentation tests are used to assess microorganisms' ability to ferment specific sugars, resulting in acid and/or gas. In these tests, the microorganism is added to a broth containing the carbohydrate of interest and a pH indicator.

Acid production causes a color change in the pH indicator, whereas gas production can be detected with a Durham tube. These tests are especially useful for distinguishing between Enterobacteriaceae species, which ferment different carbohydrates (Da Silva et al., 2018).

2.5.2. Catalase Test

The catalase test detects the presence of catalase, an enzyme that converts hydrogen peroxide into water and oxygen. When a microbial colony is exposed to hydrogen peroxide, the formation of bubbles indicates a positive catalase reaction. This test is critical for distinguishing between catalase-positive organisms such as Staphylococcus species and catalase-negative organisms such as Streptococcus species (Wieser et al., 2012).

2.5.3. Oxidase Test

The oxidase test detects bacteria that produce cytochrome C oxidase, an enzyme in the electron transport chain (ETC). When a specific reagent is applied to the bacterial colony, the color changes to dark blue or purple. This indicates a positive test. This test is essential for distinguishing oxidase-positive organisms like Pseudomonas species from oxidase-negative organisms like Enterobacteriaceae (Wieser et al., 2012).

2.5.4. Urease Test

The urease test measures microorganisms' ability to hydrolyze urea into ammonia and carbon dioxide using the enzyme urease. Ammonia production raises the medium's pH, causing the pH indicator to change color. This test is especially useful for detecting urease-positive bacteria, such as Proteus species, which can quickly hydrolyze urea (Atlas, 2010).

2.5.5. Indole Test

The indole test evaluates bacteria's ability to decompose tryptophan into indole, pyruvate, and ammonia using the enzyme tryptophanase. After incubation, the presence of indole is detected by adding Kovac's reagent, which produces a red compound. This test is useful for distinguishing indole-positive bacteria such as *Escherichia coli* from indole-negative bacteria (Wieser et al., 2012).

2.5.6. Methyl Red and Voges-Proskauer Tests

The IMViC series includes the Methyl Red (MR) and Voges-Proskauer (VP) tests for differentiating Enterobacteriaceae strains. The MR test detects mixed acid fermentation, which lowers pH and causes a red color change when methyl red is added. The VP test identifies organisms that produce acetoin during glucose fermentation, which causes a red color change when specific reagents are added. These tests aid in the distinction between MR-positive bacteria, such as *Escherichia coli* and VP-positive bacteria such as Enterobacter species (Atlas, 2010; Da Silva et al., 2018).

2.5.7. Citrate Utilization Test

The citrate utilization test measures microorganisms' ability to use citrate as their sole carbon source. The test medium contains citrate and a pH indicator. The use of citrate produces an alkaline reaction, causing the pH indicator to change color. This test is required to distinguish citrate-positive bacteria, such as Klebsiella and Enterobacter species, from citrate-negative bacteria (Da Silva et al., 2018).

2.5.8. Gelatin Hydrolysis Test

The gelatin hydrolysis test evaluates microorganisms' ability to produce gelatinases, which are enzymes that hydrolyze gelatin. After inoculation and incubation, the medium is chilled to determine whether it remains liquid, which indicates gelatin hydrolysis. This test aids in identifying gelatinase-producing bacteria, such as certain Pseudomonas and Clostridium species (Wieser et al., 2012).

2.5.9. Triple Sugar Iron Agar Test

The TSI agar test measures a microorganism's ability to ferment glucose, lactose, and sucrose and produce hydrogen sulfide. The medium contains these sugars, phenol red to indicate pH, and ferrous sulfate to detect hydrogen sulfide production. Changes in the slant and butt of the agar, as well as blackening caused by hydrogen sulfide, provide a complete picture of the organism's fermentation capabilities and gas production. This test is commonly used to identify Enterobacteriaceae (Da Silva et al., 2018; Wieser et al., 2012).

2.6. MOLECULAR TECHNIQUES IN MICROBIOLOGY

Molecular methods are critical tools for analyzing microorganisms found in food and other biological components. The procedures allow for the screening of a wide range of compounds in a single test. It has genuinely matured, and its scope of use is expected to expand in the near future. The latter method has been adopted by the food industry, water processors, and analytical laboratories for quick species distinction, strain identification, and strain-relatedness determination from infected samples. Molecular approaches differ in terms of discriminatory power, repeatability, usability, and interpretability.

Below are some of the most commonly used molecular techniques in microbiology, each with a specialized application:

2.6.1. Polymerase Chain Reaction (PCR)

PCR is a key technique in molecular biology that amplifies specific DNA sequences. This process comprises multiple cycles of denaturation, annealing, and extension, which are aided by DNA polymerase. PCR detects and analyzes small amounts of DNA, making it essential for diagnostics, cloning, and gene expression investigations. The specificity and sensitivity of this tool have transformed molecular diagnostics and research, enabling quick identification of infections and genetic alterations (Jinek et al., 2012).

2.6.2. Gel Electrophoresis

Microbial quality control is an important part of microbiology that ensures product safety and dependability in a variety of industries. The application of contemporary molecular techniques has improved the accuracy and efficiency of microbiological QC. However, challenges like as recognizing VBNC organisms, standardizing techniques, and understanding complex data remain. Continued research and development in this sector are required to overcome these issues and improve the effectiveness of microbiological quality control systems (Lederberg and Relman, 2009).

2.6.3. Sequencing Technology

Sequencing technology encompasses techniques for determining the nucleotide sequence of DNA. High-throughput genome analysis is now possible because of techniques, such as Sanger sequencing and next-generation sequencing (NGS), which allow for detailed assessments of gene variations and functions. NGS has transformed microbiology by giving comprehensive data on microbial diversity, evolution, and disease. These technologies are crucial for metagenomics, epidemiology, and personalized medicine, accelerating progress in understanding complex biological systems (Mardis, 2008).

2.6.4. Quantitative PCR (qPCR)

Quantitative PCR (qPCR), also known as real-time PCR, detects DNA or RNA levels in a sample by tracking the amplification process in real-time. qPCR, which employs fluorescent dyes or probes, provides both qualitative and quantitative data, allowing for precise assessment of gene expression levels and the identification of low-abundance targets. This method is widely used in clinical diagnostics, pathogen detection, and gene expression investigations due to its sensitivity, specificity, and short turnaround time (Jinek et al., 2012).

2.6.5. Fluorescence in Situ Hybridization (FISH)

FISH is a cytogenetic approach in which fluorescent probes connect to specific DNA sequences on chromosomes. This allows for the visualization and localization of genetic information within cells, which aids in gene mapping, microbiological identification, and the detection of chromosomal anomalies. FISH is particularly useful in environmental microbiology to investigate microbial communities and in clinical diagnostics to discover genetic disorders and malignancies (Lederberg and Relman, 2009).

2.6.6. CRISPR-Cas9

Clustered regularly interspaced short palindromic repeat (CRISPR)-Cas9 is a groundbreaking genome editing technique that originated from the bacterial immune system. It enables precise, targeted alterations to DNA in live organisms. The technique employs a guide RNA to route the Cas9 nuclease to a specific DNA sequence, where it causes double-strand breaks (DSBs). These fractures can be repaired to add or remove genetic material, allowing gene knockout, correction, or insertion. CRISPR-Cas9 holds enormous potential in genetics, medicine, and biotechnology, providing unparalleled capabilities for gene therapy and functional genomics (Jinek et al., 2012).

2.6.7. Microarray Analysis

A high-throughput method for examining patterns of gene expression is microarray analysis. Gene expression levels in a sample can be simultaneously assessed by covering a solid surface with thousands of DNA probes. Because it makes it possible to identify genes that are expressed differently, better understand how genes operate, and find disease biomarkers, this technology is helpful in the field of genomics. An in-depth understanding of cellular functions is provided by microarrays, which are useful for drug development, customized treatment, and diagnostics (Lederberg and Relman, 2009).

The following is a summary of the purpose and application of the above molecular techniques in microbiology:

1. **Polymerase Chain Reaction (PCR):** The purpose is to amplify certain DNA sequences, making microorganisms easier to study and identify. Applications include detecting pathogens, characterizing genes, and studying microbial diversity.
2. **Gel Electrophoresis:** Purpose: Sorts DNA, RNA, or proteins according to size and charge. Applications include analyzing PCR results, evaluating gene expression, and detecting genetic variations.
3. **Sequencing Technologies:** Purpose: Identifies the nucleotide sequence of DNA or RNA. Applications include whole-genome sequencing, metagenomics, and transcriptomics. NGS is particularly beneficial for thorough and high-throughput analyses.
4. **Quantitative PCR (qPCR):** Purpose: detects the amount of DNA or RNA in a sample in real-time. Applications include quantifying microbial burden, monitoring gene expression, and measuring treatment efficacy.

5. **Fluorescence In Situ Hybridization (FISH):** Uses fluorescent probes to detect and locate specific nucleic acid sequences in cells or tissues. Applications include identifying and classifying microorganisms in complicated samples like environmental or clinical specimens.
6. **CRISPR-Cas9:** Enables precision genome editing by producing targeted alterations. Applications include functional genomics, genetically engineered microorganism development, and gene function research.
7. **Microarray Analysis:** Measures the expression of thousands of genes on a chip using numerous probes. Applications include gene expression profiling, finding gene expression trends, and investigating microbial responses to different circumstances (Murray et al., 2015).

These techniques are integral to modern microbiology for diagnostics, research, and understanding microbial functions and interactions.

2.7. MICROBIAL ASSAY METHODS

Microbial assays are a method for determining the concentration or potency of a substance. This is accomplished by introducing the substance to microorganisms and observing how it affects them.

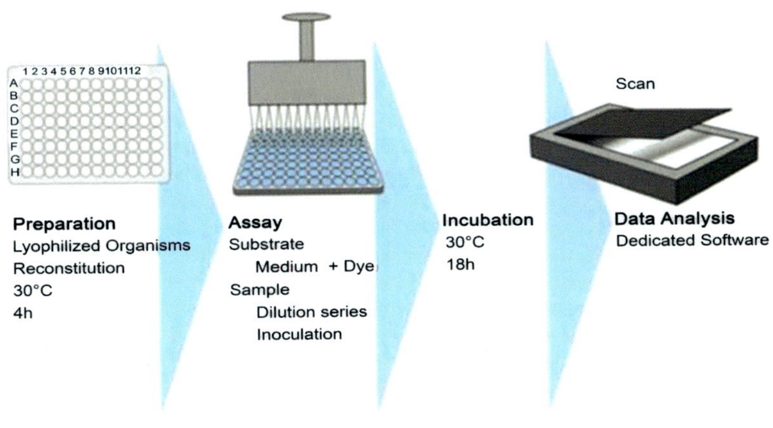

Figure 2.5. *Microbial assay for risk assessment (MARA).*

Source: Zubaidi MA, Proch J, Konieczny P, Tomczyk Ł. (2021). Toxicity testing by the microbial assay for risk assessment (MARA) in relation to trace elements content in king bolete (Boletus edulis) collected in several sites of Poland. Applied Sciences, 11(9):4166. https://doi.org/10.3390/app11094166.

The following are some of the microbial assay methods frequently in microbiology.

2.7.1. Disc Diffusion Method (Cylindrical Cup Method)

The cylindrical cup method along with various disc diffusion methods are frequently used to assess antibacterial activity. This method involves inoculating an agar surface with the test microorganism and then placing paper discs or cylindrical cups packed with antibiotics on it. A concentration gradient is produced by the antibiotic's radial diffusion. The antibiotic's effectiveness is indicated by measuring the region surrounding the disc or cup where bacteria are unable to proliferate. This approach is easy to use, economical, and enables the testing of several compounds at once. Antimicrobial susceptibility testing has adopted it as a standard due to its dependability and repeatability. Agar thickness and the difficulty of directly measuring the minimum inhibitory concentration (MIC) are two limitations. Variations of this method have been extensively applied in screening the antimicrobial properties of medicinal plants (Hewitt, 2004).

2.7.2. Tube Assay (Turbidimetric Method)

The tube test, also known as the turbidimetric method, determines a substance's antimicrobial activity by measuring the turbidity of a microbial culture. The approach is based on reducing turbidity by inhibiting microbial growth in the presence of an antibiotic drug. This approach includes mixing the test substance with a microbiological suspension in a test tube and measuring the optical density (OD) at predetermined intervals. The decrease in turbidity suggests that the antibacterial medication is effective. This approach enables the continuous monitoring and quantification of microbial growth. However, it necessitates accurate instrumentation and is susceptible to changes in experimental conditions. According to a study by Hewitt (2004), the turbidimetric method is a critical tool in assessing the antibacterial activity in various natural compounds.

2.7.3. Urease Assay

The urease assay is a biochemical test that detects the presence of the urease enzyme, which catalyzes the breakdown of urea into ammonia and carbon dioxide. This technique is useful in microbiology for detecting urease-positive bacteria, such as Helicobacter pylori. The test consists of incubation of the bacterial culture with a urea-containing medium and a pH indicator. Ammonia production elevates the pH, resulting in a color shift in the medium. The urease assay is quick, easy, and specific, making it a useful diagnostic tool. However, it is limited to urease-producing organisms and requires careful interpretation.

According to research by Snaith et al. (2014), the urease assay remains a cornerstone in the detection of *H. pylori* in clinical samples (Snaith et al, 2014).

2.7.4. Luciferase Assay

The luciferase enzyme catalyzes the oxidation of luciferin, and the luciferase assay is a highly sensitive technique for measuring the bioluminescence emitted by this enzyme. This assay is commonly used to investigate gene expression, microbial contamination, and cellular ATP levels in molecular biology, microbiology, and environmental monitoring. A luminometer is used to measure the light that is produced by the luciferase reaction. This method's excellent sensitivity, specificity, and quick findings make it valuable. It frequently serves as an instrument for high-throughput screening and enables real-time biological process monitoring. Some restrictions include the need for particular substrates and the possibility of interference from other luminous materials. Therefore, the luciferase assay has been instrumental in advancing our understanding of microbial interactions and environmental microbiology (Hewitt, 2004).

2.8. MICROBIAL PRESERVATION TECHNIQUES

To keep microorganisms viable and stable for use in science, industry, and medicine, microbial preservation is crucial. Over the past few decades, microbial preservation techniques have undergone tremendous evolution, with breakthroughs boosting these technologies' sustainability, cost-effectiveness, and efficiency. Here, we summarize the main methods of microbial preservation that have been used and developed, emphasizing important research findings and recent advancements in the field.

2.8.1. Cryopreservation

Cryopreservation involves storing microorganisms at ultra-low temperatures, typically in liquid nitrogen at -196°C or in deep freezers at -80°C. This method is widely regarded for its effectiveness in preserving a wide range of microbial species with minimal genetic and phenotypic changes over time. Studies have shown that cryopreservation can maintain microbial viability and metabolic activity for extended periods (Prakash et al., 2020). The use of cry protectants, such as glycerol and dimethyl sulfoxide (DMSO), is crucial to prevent ice crystal formation, which can damage cell membranes (Mañas and Pagán, 2005).

2.8.2. Lyophilization (Freeze-Drying)

Another popular method is lyophilization, which is also known as freeze-drying, which is freezing microbial cultures and then sublimating ice under a vacuum. A dry, stable product that can be kept at room temperature is the end result of this method. Because lyophilization requires less space and is simple to reconstitute, it is especially beneficial for the long-term storage and transportation of microbial cultures (De Vero et al., 2019). Studies have demonstrated the efficiency of lyophilization in preserving the stability and viability of yeasts, fungi, and bacteria (Alonso, 2016).

2.8.3. Encapsulation

Encapsulation encloses microbial cells in a protective matrix, such as alginate, gelatin, or polyvinyl alcohol, which protects them from environmental stresses. This approach is especially useful for keeping probiotics and other beneficial bacteria found in the food and pharmaceutical industries. Encapsulation increases the stability and controlled release of microbial cells, extending their shelf life and efficacy (Mañas and Pagán, 2005).

2.8.4. Spray Drying

Spray drying is a method of dehydration that uses hot gas to rapidly dry liquid microbial cultures into a powder. This approach is both efficient and cost-effective, making it ideal for large-scale industrial applications. Spray drying has been found to sustain microbial viability, and is routinely employed to preserve bacterial cultures used in agricultural and food production (Alonso, 2016).

2.8.5. Overlay of Oil

Using the oil overlay approach, sterile mineral oil is applied to microbiological cultures. This method works well for keeping microorganisms that are susceptible to oxidative stress and desiccation. Anaerobic environments produced by oil overlay are advantageous for the maintenance of anaerobic microbes. Research has shown that this technique can sustain culture viability for a number of years (Alonso, 2016).

2.8.6. Extra-Cold Freezing Temperature

Microbial cultures are frequently preserved for short to medium periods of time via ultra-low temperature freezing, which is usually done at -80°C. It is a rather easy and affordable method that can sustain microbial viability for a few months to a few years. To avoid cellular damage, repetitive freeze-thaw cycles should be avoided (Manas and Pagán, 2005).

2.8.7. High Hydrostatic Pressure

High hydrostatic pressure (HHP) is an emerging technique that applies extreme pressure to microbial cultures, which can inactivate spoilage organisms and pathogens while preserving the viability of desired microorganisms. HHP is particularly useful in food preservation, where it is employed to extend shelf life and enhance food safety without the use of heat or chemicals (Alonso, 2016).

2.8.8. Anhydrobiosis

Anhydrobiosis, also known as dry state preservation, is the process of drying out microorganisms to the point where their metabolism stops but their cells are still alive. The ability of some extremophiles to endure severe dehydration served as the model for this procedure. For the preservation of bacterial spores and certain fungi that can be reactivated by rehydrating, anhydrobiosis is especially useful (Manas and Pagán, 2005).

2.9. MICROBIAL QUALITY CONTROL

Microbial quality control (QC) is an important area of microbiology that assures product safety, efficacy, and reliability, especially in industries like pharmaceuticals, food processing, and healthcare. This technique entails a series of systematic procedures for monitoring and controlling the presence and concentration of microorganisms. It aids in preventing contamination, assuring product safety, and adhering to regulatory requirements. Microbial quality control has grown in relevance as molecular tools have advanced and our understanding of microbial communities has improved. This debate delves into the importance, procedures, and issues related to microbiological quality control.

2.9.1. Significance of Microbial Quality Control

Microbial QC is critical to public health and safety. For example, in the food business, microbial contamination can result in foodborne illnesses, which are a serious public health concern. According to Jasson et al. (2010), the use of microbiological quality control procedures such as Hazard Analysis Critical Control Points (HACCP) has dramatically reduced the prevalence of foodborne infections. HACCP entails identifying important points in the manufacturing process where contamination may occur and adopting controls to reduce these risks. This systematic methodology assures that food products are safe to consume.

In the pharmaceutical sector, microbial quality control is critical to ensuring that goods are free of dangerous germs. Köser et al. (2012) emphasize the value of whole genome sequencing (WGS) in diagnostic and public health microbiology. WGS offers a comprehensive approach to finding and characterizing microbiological contaminants, resulting in more accurate and reliable QC measurements. This technology has become a typical tool in modern diagnostic microbiology laboratories, allowing for more accurate microbial contamination control and prevention (Jasson et al., 2010).

2.9.2. Methodologies in Microbial Quality Control

Microbial QC employs a variety of procedures, including traditional culture-based methods and sophisticated molecular techniques. Traditional approaches, such as using culture media to cultivate and identify microorganisms, remain a cornerstone of microbiological quality control. However, these approaches are time-consuming and may not detect all bacteria.

Advances in molecular approaches have transformed microbiological quality control. Microorganisms can be detected quickly and precisely using techniques including PCR, qPCR, and NGS. qPCR has emerged as a significant tool in microbial diagnostics due to its capacity to measure microbial burden and detect specific diseases. This technology has been widely implemented in a variety of industries due to its high accuracy and quickness in identifying microbiological contamination.

Another significant advancement is metagenomic analysis, which allows for the comprehensive study of microbial communities in a given environment (Jasson et al., 2010), discuss the importance of quality control and preprocessing in metagenomic datasets to ensure accurate analysis. These techniques have enhanced the ability to monitor microbial populations and identify potential contaminants in various settings, from clinical diagnostics to environmental monitoring.

2.9.3. Challenges in Microbial Quality Control

Despite advances in microbiological quality control approaches, a number of difficulties persist. One key problem is detecting viable but non-culturable bacteria (VBNC). These are bacteria that are alive but do not grow on ordinary culture media, making them difficult to detect with conventional methods. Molecular tools such as qPCR and NGS can help overcome this difficulty by detecting these organisms' genetic material, but they may not always distinguish between living and dead cells.

Another problem is the standardization and validation of new quality control procedures. The Microbiome Quality Control (MBQC) Project Consortium (2017) emphasizes the importance of defined methodologies and reference materials in ensuring the reliability and reproducibility of microbial QC results. This is especially significant in approaches like metagenomics, where differences in sample processing and data analysis can result in varied conclusions.

The interpretation of microbiological QC data might be difficult, requiring particular knowledge and skills. The diversity of microbial communities, as well as the dynamic nature of microbial populations, contribute to this complexity. Effective quality control measures must account for these elements to give accurate and actionable results (Mañas and Pagán, 2005).

2.10. MICROBIAL SAFETY PROCEDURES

Microbial safety procedures are critical in microbiology to prevent contamination, ensure the reliability of experimental results, and protect human health. These procedures encompass various practices, guidelines, and regulatory frameworks designed to minimize the risk of microbial contamination in laboratory settings and food production environments. The following discussion explores microbial safety procedures based on recent scholarly literature.

2.10.1. Laboratory Safety Procedures

Microbiology laboratory safety requires strict measures to prevent contamination and ensure personnel safety. The "Clinical Microbiology Procedures Handbook" (2020) is a comprehensive resource on this topic, providing extensive step-by-step methods for treating clinical specimens and performing microbiological tests. The manual highlights the need for personal protective equipment (PPE), such as gloves, lab coats, and face shields, to safeguard laboratory personnel from harmful microbes. Furthermore, it argues for the use of biological safety cabinets (BSCs) for handling hazardous chemicals, careful disposal of biohazardous waste, and frequent training for laboratory personnel in biosafety measures (Leber, 2020).

2.10.2. Food Microbiology Safety

In food microbiology, guaranteeing the microbiological safety of food items is critical for public health. Predictive microbiology has developed as an important method for identifying and managing microbial concerns in food. According to McMeekin et al. (2002), predictive microbiology uses mathematical models to forecast pathogen growth, survival, and inactivation under varied environmental conditions. This technique helps to build effective food safety management systems (FSMS) by identifying critical control points (CCPs) where interventions can be implemented to prevent microbiological contamination. For example, managing temperature, pH, and water activity can dramatically lower the danger of microbial growth in food products (Lederberg and Relman, 2009).

2.10.3. Regulatory Frameworks and Compliance

Regulatory frameworks play an important role in ensuring microbiological safety in food production and laboratories. The Hazard Analysis and Critical Control Points (HACCP) system is a widely accepted approach to controlling food safety concerns. It entails identifying potential hazards, selecting CCPs, and putting in place procedures to limit these risks. A microbiological assessment scheme has been by researchers to evaluate the effectiveness of FSMS. This method assesses the efficacy of microbiological controls as well as conformity with food safety objectives. The authors underline that regular monitoring and verification are needed to guarantee that the adopted safety measures are effective and comply with regulatory criteria (Murray et al., 2015).

2.10.4. Advanced Microbial Safety Techniques

Novel approaches to microbiological safety have been devised to improve pathogen identification and management. HHP processing is one method for rendering microorganisms in food products inactive. According to Rendueles et al. (2011), who reviewed the use of HHP in food safety, it successfully lowers microbial burdens without sacrificing the product's sensory appeal. This non-thermal technique is an important tool for guaranteeing the microbiological safety of perishable food products since it breaks down the cellular structure of infections (Murray et al., 2015).

2.10.5. Future Challenges and Research Directions

Future concerns persist even with the developments in microbiological safety practices. Mañas and Pagán (2005) emphasized that the introduction of novel diseases and the complexity of food supply chains call for a worldwide approach to microbiological food safety. To overcome these issues, the authors support global cooperation and the creation of standardized standards. Furthermore, they stress the significance of the study in comprehending the mechanisms of microbial resistance and the creation of novel therapies to improve microbiological safety (Mañas and Pagán, 2005).

CHAPTER 3

MICROBIAL PHYSIOLOGY

Contents

Microbial physiology studies vital processes in microorganisms such as bacteria, fungi, algae, and archaea. Moat et al. (2002) investigate the metabolic pathways, genetic regulation, and cellular structures that allow these organisms to thrive in diverse environments. Understanding microbial physiology is critical for many fields, including biotechnology, medicine, and environmental science. For example, understanding microbial metabolism can help advance antibiotic development and optimize industrial fermentation processes.

Furthermore, microbial physiology is important in ecological studies because it explains microorganisms' roles in biogeochemical cycles and how they respond to environmental changes. Researchers continue to uncover the complex mechanisms that drive microbial life by combining techniques from molecular biology, biochemistry, and genetics, resulting in innovations in health, industry, and environmental management (Madigan et al., 2018).

3.1. BACTERIAL PHYSIOLOGY

Bacterial physiology refers to the biochemical and genetic mechanisms that control bacterial growth, metabolism, and regulation. These single-celled organisms are remarkably adaptable, allowing them to thrive in a wide variety of environments.

3.1.1. Growth and Reproduction

Bacterial growth typically occurs in four phases: lag, log (exponential), stationary, and death.

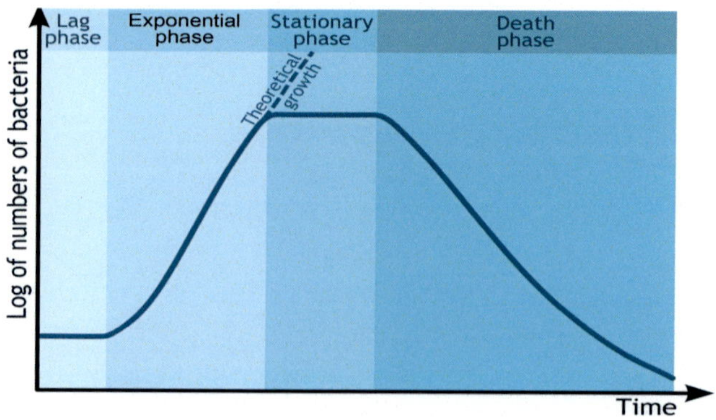

Figure 3.1. *Bacterial growth curve.*

Source: https://bio.libretexts.org/Bookshelves/Microbiology/Microbiology_%28Boundless%29/06%3A_Culturing_Microorganisms/6.06%3A_Microbial_Growth/6.6E%3A_Generation_Time.

During the lag phase, bacteria adapt to their surroundings by producing necessary enzymes and molecules. Rapid cell division characterizes the log phase, during which the population doubles at a constant rate. Optimal conditions, such as nutrient availability and favorable temperature and pH levels, aid in this exponential growth (Madigan et al., 2018). During the stationary phase, nutrient depletion and waste accumulation halt growth, resulting in a dynamic equilibrium. When adverse conditions prevail, cell lysis occurs, and the population declines.

3.1.2. Metabolism

Catabolism and anabolism are the two types of bacterial metabolism. Catabolic pathways degrade organic and inorganic substrates to produce energy, primarily through glycolysis, the tricarboxylic acid (TCA) cycle, and oxidative phosphorylation. Glycolysis converts glucose to pyruvate, which produces ATP and NADH. Pyruvate enters the TCA cycle, producing more ATP, NADH, and FADH2. These electron carriers drive oxidative phosphorylation in the electron transport chain (ETC), which generates significant ATP via chemiosmosis (Moat et al., 2002).

Bacteria use a variety of metabolic strategies, including aerobic and anaerobic respiration, as well as fermentation. Aerobic respiration uses oxygen as the terminal electron acceptor, whereas anaerobic respiration uses other acceptors such as nitrate, sulfate, or CO_2. Fermentation is an anaerobic process that produces ATP via substrate-level phosphorylation while also producing organic byproducts like lactic acid or ethanol (Murray et al., 2015).

Anabolic pathways use ATP and NADPH to synthesize cellular components. The Calvin cycle is an important process for carbon fixation, amino acid biosynthesis, and nucleotide synthesis. The balance of catabolism and anabolism is critical for bacterial growth and survival, and it is controlled by complex feedback mechanisms and environmental cues (Madigan et al., 2018).

3.1.3. Genetic Regulation

Oprons, regulatory proteins, and small RNAs all play a role in bacterial genetic regulation. The lac operon in *Escherichia coli* serves as an example of gene regulation in response to environmental changes (Moat et al., 2002). Lactose activates the lac operon, allowing genes required for lactose metabolism to be transcribed. This process requires the interaction of repressor proteins, inducers, and promoter sequences.

Another key regulatory mechanism is quorum sensing, which allows bacteria to coordinate their behavior based on population density. This cell-to-cell communication involves the production and detection of signaling molecules known as autoinducers. When a critical concentration is reached, autoinducers bind to receptor proteins and activate the transcription of target genes involved in bioluminescence, virulence, and biofilm formation (Murray et al., 2015).

Bacteria's two-component regulatory systems allow them to detect and respond to environmental stimuli. These systems include a sensor kinase and a response regulator. When the sensor kinase detects a specific signal, it phosphorylates the response regulator, changing its activity and modulating gene expression accordingly. This mechanism is critical for bacterial adaptation and survival in dynamic environments (Moat et al., 2002).

3.2. FUNGAL PHYSIOLOGY

Fungal physiology refers to the biochemical, structural, and regulatory processes that help fungi grow, survive, and reproduce. Understanding fungal physiology provides information about their metabolic pathways, cellular structures, and environmental interactions.

3.2.1 Cell Structure and Function

Fungi have a distinct cellular architecture consisting of a rigid cell wall, a cell membrane, and complex internal organelles. The fungal cell wall, which is primarily made up of chitin, glucans, and proteins, provides structural integrity and protection from environmental stresses (Moat et al., 2002).

Chitin, a polymer of N-acetylglucosamine, is an essential component that provides rigidity and resistance to osmotic pressure. The cell membrane contains ergosterol, a sterol found only in fungi that is essential for membrane fluidity and integrity (Allison et al., 2010).

Fungal growth is typically characterized by the extension of hyphae, which are tubular structures that form an interconnected network known as mycelium. The Spitzenkörper, a vesicle supply center at the hyphal tip, drives polarized hyphal growth by directing cell wall synthesis and membrane expansion. This mode of growth enables fungi to efficiently colonize substrates and utilize nutrient sources (Anderson, 2003).

3.2.2. Metabolism

Fungal metabolism includes a diverse set of biochemical pathways, allowing them to thrive in a variety of environments. The primary metabolic pathways are glycolysis, the TCA cycle, and oxidative phosphorylation. Glycolysis converts glucose into pyruvate, which produces ATP and NADH. The TCA cycle further metabolizes pyruvate, producing more ATP, NADH, and FADH2. These electron carriers drive oxidative phosphorylation in mitochondria, which generates ATP via chemiosmosis (Moat et al., 2002; Allison et al., 2010).

Fungi are metabolically versatile, and capable of aerobic respiration, anaerobic respiration, and fermentation. Fungi can generate ATP under anaerobic conditions by fermenting ethanol, lactic acid, or other organic byproducts. Secondary metabolism in fungi produces a variety of compounds, including antibiotics (e.g., penicillin), mycotoxins (e.g., aflatoxins), and pigments. These secondary metabolites frequently serve ecological functions, such as deterring competitors or facilitating symbiotic relationships (Anderson, 2003).

3.2.3. Genetic Regulation

Fungal genetic regulation entails intricate networks of genes, regulatory proteins, and signaling pathways. Fungi have tightly regulated gene expression in response to environmental cues, allowing them to adapt to changing conditions. Transcription factors, small RNAs, and epigenetic modifications work together to regulate metabolic pathways, stress responses, and developmental processes (Moat et al., 2002).

Quorum sensing, a cell-to-cell communication mechanism, enables fungi to coordinate group behaviors like biofilm formation and virulence. This process entails the production and detection of signaling molecules, which accumulate in the environment and cause gene expression changes when a certain concentration is reached (Anderson, 2003).

Fungal regulatory systems consist of two components that allow them to sense and respond to environmental stimuli. These systems include a sensor kinase and a response regulator. When the sensor kinase detects a specific signal, it phosphorylates the response regulator, modulating its activity and altering gene expression. This mechanism is critical for fungal adaptation and survival in dynamic environments (Moat et al., 2002).

3.2.4. Nutrient Acquisition

Fungi are heterotrophic organisms that obtain nutrients from the absorption of organic matter. Saprophytic fungi decompose dead organic material and play an important role in nutrient cycling and ecosystem function. They secrete extracellular enzymes, such as cellulases, ligninases, and proteases, to break down complex polymers into simpler compounds that can be absorbed (Madigan et al., 2018).

Parasitic fungi obtain nutrients from living hosts and frequently cause diseases in plants, animals, and humans. These fungi have developed sophisticated mechanisms to invade host tissues, avoid immune responses, and extract nutrients (Moat et al., 2002). Symbiotic fungi, including mycorrhizal fungi, form mutualistic relationships with plants, facilitating nutrient exchange while also improving plant growth and resilience.

3.2.5. Environmental Adaptations

Fungi have numerous adaptations to environmental stresses, including temperature extremes, oxidative stress, and osmotic pressure. Heat shock proteins (HSPs) and antioxidant enzymes protect fungal cells from damage and maintain cellular homeostasis in stressful situations. Morphological changes, such as the transition from hyphal to yeast forms in dimorphic fungi, allow for adaptation to host environments while increasing pathogenicity (Moat et al., 2002).

3.3. VIRAL PHYSIOLOGY

Viruses, unlike cellular organisms, lack metabolic machinery and cellular structures, making them obligate intracellular parasites that rely solely on host cells for reproduction. Understanding viral physiology is critical for developing antiviral therapies, vaccines, and strategies to combat viral infections.

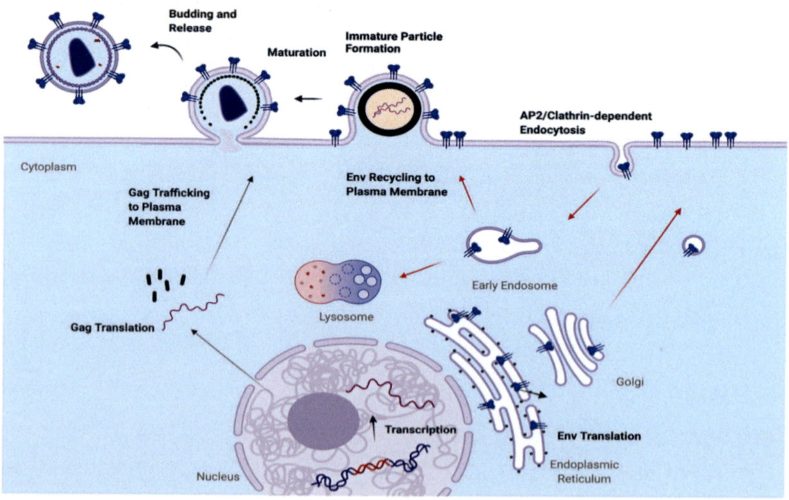

Figure 3.2. Intracellular trafficking pathways and sites of membrane alterations that are induced by different viruses.

Source: Anokhin B, Spearman P. (2022). Viral and host factors regulating hiv-1 envelope protein trafficking and particle incorporation. Viruses, 14(8), 1729. https://doi.org/10.3390/v14081729.

3.3.1. Cellular Morphology and Physiological Processes

Viruses are unique in their simplicity, with nucleic acid encased in a protein coat known as a capsid. Their genomes, made of DNA or RNA, vary in size and complexity, encoding essential proteins for replication and host manipulation. The capsid is made up of protein subunits known as capsomers, which protects the viral genome while also facilitating host cell entry. Some viruses have a lipid envelope derived from the host cell membrane that contains glycoproteins necessary for binding to host receptors and facilitating entry. This simple structure enables viruses to successfully hijack host cellular machinery, ensuring their spread and survival within host organisms (Moat et al., 2002).

3.3.2. Metabolism

Viruses lack intrinsic metabolic machinery and are metabolically inactive outside of their host. They are entirely dependent on host cells to generate energy and synthesize biomolecules. Once inside a host cell, viruses take over the host's replication and protein synthesis machinery. Viral genome replication strategies differ, with DNA viruses using host polymerases and RNA viruses frequently carrying their own polymerases. Host ribosomes translate viral mRNA into viral proteins, which are required for the assembly of new virions. This parasitic reliance on host metabolism highlights the complex interaction between viruses and their hosts, prompting the development of antiviral strategies to disrupt these processes (Moat et al., 2002; Murray et al., 2015).

3.3.3. Genetic Regulation

Viral genetic regulation ensures the efficient expression of viral genes, which is required for replication and assembly. Viral gene expression is time-dependent, with early genes encoding replication and host-modifying proteins and late genes encoding structural proteins. This regulation promotes replication efficiency and virion assembly. Viruses alter host transcription and translation to promote viral protein synthesis, frequently by degrading host mRNA or inhibiting host protein synthesis. High mutation rates, particularly in RNA viruses, and mechanisms such as recombination contribute to genetic diversity, allowing for rapid adaptation to host defenses and antiviral treatments (Maier et al., 2009; Allison et al., 2010).

3.3.4. Nutrient Acquisition

Viruses do not acquire nutrients in the traditional way but instead use host cellular resources to meet their needs. They rely on the host cell's nucleotide pools for genome replication and frequently stimulate nucleotide biosynthesis to ensure a steady supply. Amino acids from the host are used to synthesize viral proteins, while ATP produced by the host's metabolic processes powers various stages of the viral life cycle, such as genome replication and virion assembly. This reliance on host-derived nutrients emphasizes viruses' parasitic nature and suggests potential targets for antiviral strategies aimed at disrupting these critical host-virus interactions (Maier et al., 2009).

3.3.5. Environmental Adaptations

Viruses have evolved a variety of adaptations to survive and persist in a wide range of environments. Non-enveloped viruses are generally more resistant to environmental stresses such as desiccation, acidic conditions, and detergents, allowing them to survive for extended periods of time outside of a host. Viruses use strategies to avoid host immune responses, such as rapid mutation, immune inhibition, and latency, in which the virus remains dormant within host cells to avoid detection.

Furthermore, viruses can adapt to new host species via mutations that alter receptor binding specificity, allowing for cross-species transmission and the emergence of new diseases, demonstrating their evolutionary adaptability (Maier et al., 2009; Murray et al., 2015).

3.4. PROTOZOAN PHYSIOLOGY

Protozoa are a group of unicellular eukaryotic microorganisms that perform a variety of physiological functions. These organisms live in a variety of habitats, including freshwater, marine environments, and soil, and can be free-living or parasitic. Key physiological processes in protozoa include:

3.4.1. Cellular Morphology and Physiological Processes

Protozoa are unicellular eukaryotic organisms with a complex cellular structure that includes a distinct nucleus and various organelles. The cell membrane, which consists of a lipid bilayer with embedded proteins, aids in nutrient uptake, waste excretion, and environmental interaction. The nucleus contains the genetic material in linear chromosomes, while organelles such as mitochondria, endoplasmic reticulum, Golgi apparatus, and lysosomes carry out essential cellular functions.

Contractile vacuoles are specialized structures that maintain osmotic balance, while locomotory appendages such as flagella, cilia, and pseudopodia allow for movement and feeding. This intricate organization enables protozoa to carry out all necessary life processes within a single cell while adapting to changing environmental conditions and ecological niches (Moat et al., 2002).

3.4.2. Metabolism

Protozoa are metabolically versatile, adapting to both aerobic and anaerobic conditions. Aerobic protozoa use oxygen for cellular respiration, which includes glycolysis, the TCA cycle, and oxidative phosphorylation to generate ATP. Anaerobic protozoa use fermentation pathways to generate ATP from substrate-level phosphorylation and produce end products such as lactate or ethanol. Hydrogenosomes, found in some anaerobic protozoa, produce ATP and hydrogen gas. This metabolic diversity enables protozoa to thrive in a wide range of environments, from oxygen-rich to oxygen-deficient habitats, demonstrating their adaptability and evolutionary success in a variety of ecological contexts (Maier et al., 2009).

3.4.3. Genetic Regulation

Protozoa regulate gene expression in response to environmental changes and developmental signals. Promoter elements and transcription factors bind to specific DNA sequences to regulate transcription, ensuring the timely expression of necessary genes. Post-transcriptional mechanisms, such as RNA splicing, editing, and RNA interference (RNAi), help to refine gene expression. RNAi pathways degrade specific mRNAs, which regulate protein synthesis and enable rapid adaptation to environmental changes. These sophisticated regulatory mechanisms allow protozoa to respond dynamically to their surroundings, maximizing survival and reproductive success (Bonkowski, 2004).

3.4.4. Nutrient Acquisition

Protozoa obtain nutrients using a variety of mechanisms tailored to their diverse lifestyles. Phagocytosis is the process of engulfing particulate food and forming a phagosome that then fuses with lysosomes for digestion. Pinocytosis enables the uptake of dissolved nutrients via vesicle formation. Nutrient absorption via diffusion and active transport mechanisms across the cell membrane allows for direct uptake of small molecules such as amino acids, sugars, and ions. These processes ensure that protozoa receive essential nutrients for growth, energy production, and cellular functions, allowing them to survive in a variety of environmental conditions and ecological roles (Bonkowski, 2004; Maier et al., 2009).

3.4.5. Environmental Adaptations

Protozoa have remarkable adaptations that allow them to endure a variety of environmental conditions. In freshwater environments, osmoregulation maintains osmotic balance by expelling excess water through contractile vacuoles. When unfavorable conditions arise, encystment enables protozoa to form resistant cysts, allowing for survival until favorable conditions return. Parasitic protozoa have developed defense mechanisms against host immune responses, such as immune-modulatory molecule secretion and antigenic variation. These adaptations demonstrate the evolutionary strategies that protozoa use to survive and reproduce in a variety of ecological niches, maintain homeostasis, and thrive in fluctuating environments (Moat et al., 2002).

3.5. ALGAL PHYSIOLOGY

Algae, a diverse group of photosynthetic organisms, play important roles in aquatic ecosystems and have significant biotechnological potential. Their physiology allows them to thrive in a wide range of environments, including freshwater and marine habitats.

Figure 3.3. *Algae are primary producers, supporting aquatic ecosystems.*

Source: https://courses.lumenlearning.com/suny-microbiology/chapter/algae/#:~:text=The%20algae%20are%20autotrophic%20protists,red%20algae%20and%20green%20algae.

Understanding these processes is critical for understanding the environmental impact and applications in fields such as biofuels, food, and pharmaceuticals (Maier et al., 2009).

3.5.1. Cell Structure and Function

Algal cells, which range from unicellular to complex multicellular forms, have eukaryotic characteristics such as a defined nucleus and membrane-bound organelles. Their cell wall, made up of polysaccharides like cellulose and pectin, provides structural support and protection. Photosynthesis takes place in chloroplasts, which contain pigments such as chlorophylls, carotenoids, and phycobilins. Some algae have flagella or cilia that help them move toward light or nutrients. These structures, powered by dynein motor proteins, allow for motility and environmental adaptation. Algae's complex cellular architecture enables them to efficiently perform essential life processes, ensuring their survival and ecological roles (Moat et al., 2002; Maier et al., 2009).

3.5.2. Photosynthesis

Photosynthesis in algae converts light energy into chemical energy via light-dependent reactions and the Calvin cycle. The light-dependent reactions take place in the chloroplasts' thylakoid membranes, where pigments absorb light and generate ATP and NADPH while splitting water molecules to produce oxygen. The Calvin cycle uses ATP and NADPH to convert carbon dioxide into organic molecules in the chloroplast stroma, resulting in glucose for energy and structural components. This efficient photosynthetic process allows algae to thrive in a variety of light conditions, forming the foundation of aquatic food webs and driving primary productivity in aquatic ecosystems (Moat et al., 2002).

3.5.3. Nutrient Acquisition

Algae obtain nutrients through diffusion, active transport, and endocytosis. They use membrane-bound transport proteins to absorb macronutrients such as nitrogen, phosphorus, and potassium, as well as micronutrients like iron and magnesium, from their environment. Some algae exhibit mixotrophy, which is the combination of photosynthesis and heterotrophic nutrient acquisition via the ingestion of organic particles or dissolved organic compounds. These adaptable nutrient acquisition strategies allow algae to thrive in a variety of environments, promoting growth, reproduction, and ecological roles in nutrient cycling and primary production (Maier et al., 2009).

3.5.4. Reproduction

Algae reproduce asexually and sexually, adapting their strategies to environmental conditions. Under ideal conditions, asexual reproduction, which includes binary fission, fragmentation, and spore formation, allows for rapid population growth. Sexual reproduction involves gamete fusion, which causes genetic recombination and increased diversity, improving adaptability and survival. Sexual reproduction modes vary by species, including isogamy, anisogamy, and oogamy. These reproductive strategies promote genetic diversity and resilience, allowing algae to colonize a variety of habitats while maintaining ecological balance (Madigan et al., 2018).

3.5.5. Environmental Adaptations

Algae have evolved a variety of adaptations to help them survive in different environments. Pigment diversity enables them to capture a wide range of light wavelengths, optimizing photosynthesis under a variety of lighting conditions. Osmoregulatory mechanisms, such as the synthesis of compatible solutes and the regulation of ion transport, aid in maintaining osmotic balance in salinities that fluctuate. Some algae form cysts or spores to withstand harsh conditions such as desiccation, nutrient depletion, and extreme temperatures. These adaptations allow algae to thrive in a variety of habitats, ranging from freshwater to marine, contributing to their ecological success and resilience (Moat et al., 2002).

3.6. ARCHAEAL PHYSIOLOGY

Archaea, one of life's three fundamental branches along with Bacteria and Eukarya, is a diverse group of microorganisms with distinct physiological characteristics. Since its discovery as a distinct realm in the late twentieth century, extensive study has shed light on numerous aspects of archaeal physiology. This discussion tries to consolidate major insights about archaea's metabolic processes, structural adaptations, and environmental interactions (Moat et al., 2002).

3.6.1. Unique Cell Membrane Composition

One of the defining physiological characteristics of Archaea is its unique cell membrane composition. Unlike Bacteria and Eukarya, archaeal membranes are composed of ether-linked lipids rather than ester-linked lipids. These ether bonds are more stable at extreme temperatures and pH levels, contributing to the resilience of many archaeal species in harsh environments. Additionally, archaeal lipids often include isoprenoid chains, which can form monolayers or bilayers, providing further structural stability (Maier et al., 2009).

3.6.2. Metabolic Diversity and Adaptation

Archaea have extraordinary metabolic diversity, allowing them to flourish in a variety of settings, including harsh ones. Archaea's key metabolic processes include methanogenesis, sulfur reduction, and halophilic and thermophilic adaptations.

- **Methanogenesis:** This is a distinct type of metabolism found only in a group of Archaea known as methanogens. Methanogenesis is the process of converting carbon dioxide to methane, which is an important part of the carbon cycle. Methanogens live in anaerobic settings such as marshes, ruminant stomachs, and sediments.
- **Sulfur Reduction and Oxidation**: Many Archaea use sulfur compounds as electron donors or acceptors. For example, thermophilic archaea from the genus Sulfolobus may oxidize sulfur to sulfuric acid and thrive in acidic hot springs. In contrast, some hyperthermophilic archaea convert sulfur compounds to hydrogen sulfide, which contributes to the sulfur cycle.
- **Halophilic Adaptations**: Halophilic archaea, including those in the Halobacteriaceae family, flourish in highly saline environments such as salt flats and saline lakes. These creatures have adapted by creating suitable solutes, such as potassium chloride, to balance the osmotic pressure of their surroundings. They also have extremely acidic proteins that are stable and functioning under high-salt environments.
- **Thermophilic Adaptations**: Hyperthermophilic archaea, such as those from the genera *Pyrococcus* and *Thermococcus*, can survive at temperatures above 100°C. Their proteins have stronger ionic connections and hydrophobic cores, which improves stability at high temperatures. Furthermore, these archaea have special DNA-binding proteins that prevent their genetic information from becoming denatured under high temperatures (Maier et al., 2009).

3.6.3. Genetic and Biochemical Pathways

Archaea's genetic and metabolic processes are comparable to but also distinct from those of Bacteria and Eukarya. For example, archaeal transcription and translation methods are more similar to those seen in eukaryotes, involving sophisticated RNA polymerases and ribosomal proteins. However, Archaea have their own enzymes and pathways, such as the modified Embden-Meyerhof-Parnas glycolysis pathway, which includes stages that are not seen in other domains.

Furthermore, Archaea frequently feature specific cofactors, such as coenzyme M and coenzyme B, which are essential for methanogenesis. The presence of these specialized molecules emphasizes the various evolutionary paths that have built archaean biochemistry (Moat et al., 2002).

3.6.4. Ecological and Environmental Significance

Archaea play important roles in many biological and environmental processes. Their ability to thrive in harsh conditions makes them important players in biogeochemical cycles, especially in areas where few other creatures can survive. Methanogens, for example, play an important role in anaerobic digestion, helping to produce methane, a strong greenhouse gas. Understanding these organisms' physiology is critical for creating techniques to reduce methane emissions and using methanogens for biotechnological purposes like biofuel generation.

Furthermore, archaeal populations in marine environments help to maintain the nitrogen cycle through processes like ammonia oxidation. Ammonia-oxidizing archaea (AOA) are abundant in the ocean and play an important role in nitrification, which affects nitrogen availability and ecosystem productivity (Madigan et al., 2018).

3.7. MICROBIAL GROWTH FACTORS

Microbial growth factors are vital chemicals that control microbe proliferation and metabolic activity. These elements include nutrients, ambient circumstances, and specialized chemical signals that aid in the physiological and biochemical processes required for microbial growth. Understanding microbial growth factors is important for a variety of applications, including industrial microbiology, biotechnology, and clinical research, because it allows for the adjustment of growth conditions for helpful microorganisms while also controlling pathogenic species. This understanding informs advances in antibiotic research, fermentation technologies, and environmental microbiology.

These growth factors can be categorized into four main spectrums which are discussed below:

3.7.1. Physical Factors

Physical elements have a considerable impact on microbial growth, with temperature, pH, moisture, and light being critical. Temperature influences the enzymatic activity and membrane fluidity of microorganisms (Maier et al., 2009). Mesophiles, for example, thrive at the human body temperature (37°C), making them common in therapeutic settings. Psychrophiles, on the other hand, thrive in cold conditions, such as Arctic ice, thanks to enzyme adaptations that perform best at low temperatures. PH levels are also important; most bacteria prefer neutral pH, but acidophiles and alkaliphiles can thrive in severe pH settings. Moisture is important because it aids in the breakdown and transport of nutrients into cells. Microorganisms require water for metabolic functions, and habitats with high water activity (aw) sustain thriving microbial communities. Finally, light is required for phototrophic microorganisms, which consume. Conversely, many microorganisms are sensitive to UV light, which can damage DNA and inhibit growth. Collectively, these physical factors create environments where specific microbes can thrive, highlighting the adaptability and diversity of microbial life (Moat et al., 2002).

Highlighted below is a detailed summary of the physical factors influencing the growth of microbial:

1. **Temperature**: Optimal Range: Different microorganisms have varying optimal temperature ranges for growth. These can be classified as:
   - Psychrophiles: Thrive at temperatures between -5°C and 20°C.
   - Mesophiles: Optimal growth at 20°C to 45°C; most human pathogens are mesophiles.
   - Thermophiles: Grow best at 45°C to 80°C.
   - Hyperthermophiles: Optimal growth above 80°C.
   - Temperature affects enzymatic activities and cellular processes, with each microorganism adapted to a specific range.
2. **pH:** Acidity/Alkalinity: Most bacteria prefer neutral pH (6.5–7.5), but some can thrive in acidic (acidophiles) or alkaline (alkaliphiles) conditions.
   - pH affects enzyme activity, membrane stability, and nutrient solubility.
3. **Oxygen Levels**: Aerobes: Require oxygen for growth.
   - Anaerobes: Grow in the absence of oxygen; can be obligate (killed by oxygen), facultative (can grow with or without oxygen), or aerotolerant (indifferent to oxygen).
   - Microaerophiles: Require low levels of oxygen.
4. **Moisture**: Water is essential for microbial growth as it is a solvent for nutrients and involved in metabolic processes.
   - Water activity (aw) is a measure of water availability; most bacteria require high aw (>0.90), while fungi can grow at lower aw.

5. **Light**: Phototrophic microorganisms require light for photosynthesis.
   - Some microbes are sensitive to light and prefer dark conditions (Maier et al., 2009).

3.7.2. Biological Factors

Microbiological growth is significantly influenced by biological interactions such as symbiosis, competition, and antagonism. The gut microbiota and its host have symbiotic connections that improve nutrition absorption and have metabolic advantages. As an illustration, some gut bacteria manufacture vital vitamins that the host is unable to make. Microbial development is also influenced by competition for resources, such as nutrients and available space. Microorganisms use methods, such as quicker growth rates or more effective nutrient uptake mechanisms, to outcompete rivals in competitive situations. Microorganisms frequently engage in antagonistic interactions, in which they create chemicals that hinder or eliminate rivals. This is demonstrated by the synthesis of antibiotics by some bacteria, including *Streptomyces*, which inhibit the growth of vulnerable microorganisms and hence lessen competition. Additionally, microbial communities can influence each other through quorum sensing, where chemical signaling coordinates activities like biofilm formation and virulence factor production. These biological factors create dynamic ecosystems where microbial growth is constantly shaped by interactions with other organisms, emphasizing the complexity and interdependence of microbial communities (Madigan et al., 2018).

Below is a detailed summary of biological factors that affect microbial thriving in their community:

1. **Microbial Interactions**: Symbiosis: Mutualistic relationships can enhance growth, as seen in gut flora.
   - **Competition:** Microorganisms compete for nutrients and space, influencing growth dynamics.
   - **Antagonism:** The production of antimicrobial substances (e.g., antibiotics) by some microorganisms can inhibit others.
2. **Host Factors**: For pathogens, the immune status of the host, availability of nutrients, and presence of competing microbiota affect microbial growth and colonization (Madigan et al., 2018).

3.7.3. Chemical Factors

Microbial growth depends on an array of chemical variables, such as the availability of nutrients, osmotic pressure, and the presence of growth agents. Building blocks for cellular architecture and energy synthesis are nutrients. Nitrogen is necessary for the synthesis of amino acids and nucleotides, whereas carbon sources like glucose are broken down to create ATP. Proteins and nucleic acids are examples of essential biomolecules that require the synthesis of sulfur and phosphorus. Iron and magnesium are examples of trace elements that function as cofactors in enzyme processes. Osmotic pressure affects nutrient intake and cell turgor, which in turn affects microbial growth. For example, halophiles, which maintain high intracellular concentrations of suitable solutes to neutralize osmotic pressure, flourish in high-salt environments. Growth factors, such as vitamins and amino acids, are also critical, particularly for fastidious organisms that cannot synthesize these compounds. The availability of these chemical factors determines the metabolic capabilities and growth rates of microorganisms, underlining the importance of a well-balanced chemical environment for optimal microbial proliferation (Moat et al., 2002).

The following is a summary of the chemical factors that influence microbial growth:

1. **Nutrient Availability**: Carbon Source: Needed for cellular structures and energy. Autotrophs use CO_2, while heterotrophs require organic carbon.
   - **Nitrogen Source:** Essential for amino acids, nucleotides, and other cellular components.
   - **Sulfur and Phosphorus:** Required for amino acids, vitamins, and nucleic acids.
   - **Trace Elements:** Iron, magnesium, calcium, etc., are cofactors for enzymes.
2. **Osmotic Pressure**: Microorganisms need a balance of solutes inside and outside their cells. Extreme osmotic conditions (high salt or sugar) can inhibit growth.
   - Halophiles thrive in high salt concentrations.
3. **Growth Factors**: Vitamins, amino acids, and other organic compounds that some microorganisms cannot synthesize and must obtain from the environment (Moat et al., 2002).

3.7.4. Environmental Factors

Environmental factors such as soil composition, water availability, and air quality have a substantial impact on microbial development. Soil provides a diverse ecosystem with changing nutrient levels, moisture, and pH, all of which influence the variety and activity of soil microbial communities. Soil microorganisms are essential for nutrient cycling, organic matter decomposition, and nitrogen fixation, which promotes plant growth. Water is necessary for microbial life, and aquatic settings support a varied range of microbial species that have adapted to changing oxygen levels, nutritional availability, and light penetration. Marine bacteria, for example, contribute to global biogeochemical cycles, such as the carbon and nitrogen cycles. Humidity, temperature, and particulate matter all have an impact on the survival and spread of airborne microorganisms. Environmental factors include anthropogenic influences, such as pollution and climate change, which can alter microbial communities. Understanding these environmental influences is critical for managing microbial growth in natural and artificial settings, emphasizing the need for comprehensive studies on microbial ecology and environmental microbiology (Epstein, 2013).

Here is a detailed summary of environmental factors that affect microbial growth:

1. **Soil:** Nutrient-rich soils support diverse microbial communities, while poor soils limit growth.
2. **Water:** Aquatic environments vary in nutrient availability, oxygen levels, and light penetration, influencing microbial growth.
3. **Air:** Airborne microbes depend on humidity, temperature, and nutrient particles in the air for growth (Moat et al., 2002).

3.8. MICROBIAL ADAPTATION STRATEGIES

Microorganisms are remarkably versatile and adaptable, allowing them to live and thrive in a wide range of conditions. Natural selection, genetic recombination, horizontal gene transfer (HGT), DNA damage repair, and pleiotropy all contribute to these adaptive qualities. Each of these systems contributes significantly to microbes' evolutionary success, allowing them to overcome environmental barriers and optimize their survival tactics (Madigan et al., 2018).

3.8.1. Natural Selection

Natural selection is a fundamental evolutionary mechanism in which organisms that are better suited to their environment survive and reproduce, passing on advantageous features to their offspring. This process is visible in microbial populations exposed to changing environmental conditions, where selecting forces favor mutations that improve survival. For example, in the presence of antibiotics, bacteria with resistance-conferring mutations are more likely to survive and reproduce, resulting in a preponderance of resistant strains (Lederberg and Relman, 2009). This selection pressure hastens the evolution of antibiotic resistance, indicating the effectiveness of natural selection in microbial adaptability.

3.8.2. Genetic Recombination

Genetic recombination is the exchange of genetic material across organisms, which results in new gene combinations. Microorganisms can undergo this process via mechanisms such as transformation, transduction, and conjugation. Recombination not only creates genetic variety but also plays an important function in DNA repair. For example, in *Escherichia coli*, recombination aids in DNA repair by enabling the exchange of damaged DNA segments with undamaged ones from a homologous chromosome (Lederberg and Relman, 2009). This method ensures genetic integrity and increases microbes' adaptive ability.

3.8.3. Horizontal Gene Transfer

The transfer of genetic material between creatures that are not parents and offspring is known as horizontal gene transfer or HGT. Microorganisms can quickly pick up additional genes through this method, including ones that give virulence, antibiotic resistance, or metabolic capacities. Three processes—transduction (transmission by a virus), conjugation (direct cell-to-cell transfer), and transformation (uptake of free DNA)—are involved in HGT. Multidrug-resistant bacterial strains are largely dispersed due to the acquisition of antibiotic-resistance genes by HGT (Anderson, 2003). Microorganisms can swiftly adjust to novel environmental obstacles due to their capacity to absorb and integrate foreign DNA.

3.8.4. DNA Damage Repair

Microorganisms are continuously exposed to substances that can damage DNA, such as chemical mutagens and UV light. Survival and preservation of genomic integrity depend on effective DNA repair systems. Prokaryotes have developed complex DNA repair mechanisms, such as recombinational, excision, and direct repair (Lederberg and Relman, 2009). For example, the homologous recombination pathway, which employs a homologous sequence as a template for precise repair, is involved in the repair of double-strand breaks (DSBs) in DNA. In addition to repairing DNA damage, this mechanism promotes genetic diversity and adaptability.

3.8.5. Pleiotropy

Pleiotropy is the phenomenon in which one gene influences several phenotypic features. This phenomenon connects various physiological systems, which can impact microbial adaptability. A coordinated response to environmental changes, for instance, could be produced by a mutation in a regulatory gene that modifies several metabolic pathways at once (Anderson, 2003). Pleiotropic effects allow microorganisms to more effectively adapt to complex environmental stimuli, which can increase their adaptability.

3.8.6. Additional Adaptation Strategies

In addition to the aforementioned mechanisms, microorganisms employ other strategies to adapt to their surroundings. These include:

1. **Mutator Phenotypes:** Mutator phenotypes are microorganisms that show higher rates of mutation. By creating a varied pool of genetic variants that natural selection can work upon, these high mutation rates can hasten the adaption process. In environments that are changing quickly, the adaptive advantages frequently exceed the disadvantages, even when mutator phenotypes might cause an accumulation of harmful mutations (Anderson, 2003).
2. **Stress Responses:** To withstand harsh environments, microorganisms have developed stress response mechanisms. For instance, when bacteria experience damage to their DNA, the SOS response is triggered, which starts the production of genes related to DNA repair and cell survival. Stress reactions offer both short-term and long-term genetic modifications, which can increase microbial resistance and adaptability (Moat et al., 2002).
3. **Metabolic Flexibility:** One of the most important adaptive features of microbes is their capacity to grow and produce energy on a variety of substrates. Microorganisms with metabolic flexibility can endure nutritional variations and take advantage of a variety of ecological niches. According to Lederberg and Relman (2009), metabolic pathways can be optimized by adaptive evolution to increase the effectiveness of resource use and energy output (Lederberg and Relman, 2009).

3.9. MICROBIAL STRESS RESPONSES

Microbial stress responses refer to the methods and techniques used by microorganisms to survive and adapt to harsh environmental conditions. These responses require complicated regulatory networks to regulate cellular damage, maintain homeostasis, and ensure life. Understanding these adaptive reactions is critical for developing knowledge in microbiology, biotechnology, and medical sciences, as it demonstrates the tenacity and adaptability of microbial life (Jasson et al., 2010).

3.9.1. pH Homeostasis

Microbial adaptation to fluctuating pH environments is crucial for survival and involves intricate mechanisms to maintain internal pH within a viable range. This process, known as pH homeostasis, allows microbes to thrive in acidic or alkaline conditions. Microbes employ various strategies such as proton pumps, ion exchange mechanisms, and buffer systems. Proton pumps like ATPase actively expel excess protons from the cytoplasm, thereby preventing acidification. Ion exchange mechanisms, such as antiporters, facilitate the exchange of intracellular H+ ions with extracellular Na+ or K+ ions, aiding in pH regulation. Additionally, buffer systems comprising weak acids and bases neutralize internal pH fluctuations. Studies show that Escherichia coli, for example, can regulate its internal pH through these mechanisms, ensuring enzyme functionality and cellular stability (Booth, 2002). This ability to maintain pH homeostasis is critical for microbial adaptation in diverse environments, from human gut flora to extreme habitats like acidic hot springs.

3.9.2. Altering of the Cell Membrane

Changes in the microbial cell membrane, which acts as an important barrier and interface with the external environment, are frequently necessary for adaptation to environmental stresses. Stress frequently causes modifications to the fluidity and composition of membranes. Bacteria may respond to cold stress by increasing the amount of unsaturated fatty acids in their membranes, which improves fluidity and preserves membrane functionality. On the other hand, heat stress might cause a rise in saturated fatty acids, which would stabilize the membrane structure. Additionally, changes in membrane protein composition can occur to enhance transport and signal transduction processes under stress conditions. Studies have highlighted that *Bacillus subtilis* adjusts its membrane lipid composition in response to osmotic stress, thereby preserving membrane integrity and functionality (Jasson et al., 2010). These membrane alterations are vital for microbial survival and adaptation, enabling them to maintain homeostasis and continue metabolic activities under various stress conditions.

3.9.3. Metabolic Regulation

Microbes respond to stresses via sophisticated metabolic regulation, resulting in optimal energy generation and resource utilization. Microbes can switch to more efficient metabolic pathways when nutrition availability is limited or oxidative stress is present, increasing survival. For example, the transition from aerobic respiration to fermentation under oxygen-limited conditions enables bacteria to continue producing ATP. Regulatory proteins and transcription factors play critical roles in this process by influencing the expression of genes involved in metabolic pathways. The strict response in bacteria, controlled by the alarmone (p)ppGpp, reprograms cellular metabolism during food restriction, prioritizing important operations and preserving resources. This metabolic flexibility is evident in various microbes, including the adaptive strategies of *Mycobacterium tuberculosis*, which modulates its metabolism to persist in the host under immune pressure (Maier et al., 2009). Thus, metabolic regulation is a key adaptive mechanism, enabling microbes to dynamically respond to environmental changes and stressors.

3.9.4. Protection and Repair of Macromolecules

Microbes have strong systems in place to safeguard and fix macromolecules, maintaining cellular integrity and functionality even in the face of adversity. Stressors like as heat, radiation, and oxidative stress can cause damage to proteins, DNA, and RNA in particular. In response, microorganisms generate chaperones and HSPs, which refold denatured proteins and stop them from aggregating. When DNA is damaged, repair processes such as homologous recombination and base excision repair are triggered. Furthermore, reactive oxygen species are neutralized by antioxidant enzymes such as catalase and superoxide dismutase, safeguarding cellular macromolecules. Research on *Deinococcus radiodurans*, an extremophilic bacterium, has shown that its exceptional resistance to radiation is due to highly efficient DNA repair systems and protective protein complexes (Jasson et al., 2010). These protective and repair mechanisms are essential for microbial resilience, enabling survival and adaptation in hostile environments.

3.9.5. Stress Priming

Pre-exposure to a mild stressor sometimes referred to as adaptive or cross-protection, strengthens microbial resistance to future severe stress. This behavior can be compared to a type of microbial "memory," in which an organism is primed for future difficulties by an initial exposure that triggers a defensive response. For instance, minor heat shock can cause the body to produce HSPs, which protect from temperatures that can be fatal. In a similar vein, exposure to antibiotics at sub-lethal concentrations can trigger stress response pathways, improving tolerance to larger dosages. The induction of generic stress response regulators, like RpoS in *Escherichia coli*, which controls the expression of a broad variety of stress response genes, supports this adaptive strategy (Bonkowski, 2004). Stress priming is a crucial survival mechanism, allowing microbes to anticipate and bette,r withstand environmental fluctuations and stressors.

3.10. MICROBIAL PATHOGENICITY

Microbial pathogenicity refers to the mechanisms and processes through which microorganisms cause disease in their hosts. The study of microbial pathogenicity encompasses various disciplines, including microbiology, immunology, and molecular biology (Hewitt, 2004). Let's review the major concepts in microbial pathogenicity, highlighting key mechanisms, host-pathogen interactions, and the influence of microbial virulence factors.

3.10.1. Mechanisms of Pathogenicity

Pathogenic microorganisms include bacteria, viruses, fungi, and parasites. Each type of pathogen employs unique strategies to establish infection, evade host defenses, and cause disease.

3.10.1.1. Bacterial Pathogenicity

Bacterial pathogens utilize various strategies to infect hosts, including adherence to host tissues, invasion of host cells, and the production of toxins.

1. **Adherence**: Bacteria bind to host cells with specific structures like pili and fimbriae. For example, *Escherichia coli* attaches to the epithelial cells of the urinary tract via its pili, which is critical for generating urinary tract infections.
2. **Invasion**: Bacteria can enter host cells using effector proteins released by specific mechanisms. *Salmonella enterica* employs a Type III secretion system to transfer proteins into host cells, boosting their uptake and survival within macrophages.
3. **Toxin production**: Bacterial toxins are classed as exotoxins or endotoxins. Exotoxins are released proteins that can cause considerable harm to host tissues. For example, *Corynebacterium diphtheriae* produces diphtheria toxin, which suppresses protein synthesis in host cells. Endotoxins, such as lipopolysaccharides (LPS) present in the outer membranes of Gram-negative bacteria, cause significant immunological responses that might result in septic shock (Hewitt, 2004).

3.10.1.2. Viral Pathogenicity

Viruses are obligatory intracellular parasites that replicate via the host cell's machinery. Viral pathogenicity is determined by their capacity to avoid immune detection and affect host cell processes.

1. **Attachment and Entry**: Viruses adhere to specific receptors on the surface of host cells. For example, the HIV virus targets CD4+ T cells by attaching to the CD4 receptor and the co-receptors CCR5 and CXCR4.
2. **Replication and Assembly**: Once inside the host cell, viruses employ the cellular machinery to reproduce their DNA and generate viral proteins. This mechanism frequently causes cell death and tissue

damage. The flu virus, for example, impairs the normal function of respiratory epithelial cells, causing respiratory symptoms.

3. **Immune Evasion**: Viruses have evolved various mechanisms to evade the host immune system. The herpes simplex virus (HSV) can establish latent infections in nerve cells, avoiding immune detection until reactivation (Bonkowski, 2004; Maier et al., 2009).

3.10.1.3. Fungal Pathogenicity

In immunocompromised people, fungi like the Candida species can cause infections. Fungi use techniques including biofilm development, enzyme secretion, and attachment to host tissues.

1. **Adhesion and Biofilm Formation**: *Candida albicans* develop complex cell communities known as biofilms, which are very resistant to antifungal therapies, and attach to mucosal surfaces.
2. **Enzyme Secretion**: Proteases and lipases, among other enzymes, are secreted by fungi and aid in the degradation of host tissues and invasion. Proteases produced by Aspergillus species play a role in the pathophysiology of aspergillosis, a dangerous lung infection (Allison et al., 2010).

3.10.2. Host-Pathogen Interactions

The interaction between pathogens and the host immune system plays a crucial role in determining the outcome of an infection. The host employs both innate and adaptive immune responses to combat pathogens.

- **Innate Immunity**: Using phagocytic cells, physical barriers, and antimicrobial substances, the innate immune system offers the initial line of defense. To cause infection, pathogens need to get past these barriers. The Staphylococcus aureus protein A attaches itself to the Fc region of antibodies, hence impeding opsonization and phagocytosis.
- **Adaptive Immunity**: T cells and B cells are activated as part of the adaptive immune system, which offers a more focused and durable response. Pathogens that impede antigen presentation and alter the host immune response, like Mycobacterium TB, can avoid adaptive immunity.

3.10.3. Microbial Virulence Factors

Pathogens create chemicals called virulence factors that boost their potential to cause disease. These include adhesins, poisons, enzymes, and immune-evasion mechanisms.

1. **Adhesins**: Molecules that aid in adhesion to host cells. For example, enteropathogenic *E. coli* produces the adhesin intimin, which allows for tight attachment to intestinal epithelial cells.
2. **Toxins**: Chemicals that harm host tissues. The cholera toxin generated by *Vibrio cholerae* produces severe diarrhea by increasing the secretion of water and electrolytes in the intestines.
3. **Enzymes**: Proteins that break down host tissues and aid in dissemination. Streptococci produce hyaluronidase, which breaks down hyaluronic acid in connective tissues, enabling bacterial spread.
4. **Immune evasion**: Mechanisms that help pathogens avoid detection and destruction by the host immune system. The capsule of *Cryptococcus neoformans* is an example, as it prevents phagocytosis and complement activation (Allison et al., 2010; Maier et al., 2009).

CHAPTER 4

MICROBIAL DIVERSITY

Contents

Microbial diversity refers to the wide variety of microorganisms found in different environments, including bacteria, archaea, viruses, fungi, and protists. This concept is important in microbiology because it has implications for ecosystem function, evolutionary biology, and biotechnology. Microbe diversity includes not only the number of species, but also genetic diversity, metabolic capabilities, and ecological roles (Matulich et al., 2013). Microbial diversity is critical to biogeochemical cycles, influencing nutrient cycling, decomposition processes, and soil and water quality (Maier et al., 2009). The complexity of the microbial community and the interactions between its members contribute to ecosystem adaptability and resilience. Furthermore, microbial diversity supports significant medical and industrial advancements by providing novel bioactive compounds and biocatalysts (Torsvik and Øvreås, 2002). Recent advances in high-throughput sequencing and metagenomics have dramatically increased our understanding of microbial diversity, revealing a plethora of previously unknown microorganisms and clarifying their roles in various environments. These technologies have highlighted the intricate and dynamic nature of microbial communities, revealing their critical role in both natural and engineered systems (Matulich et al., 2013).

4.1. BACTERIAL DIVERSITY

Bacterial diversity refers to the wide variation in bacterial species, genetic composition, metabolic capabilities, and ecological roles..

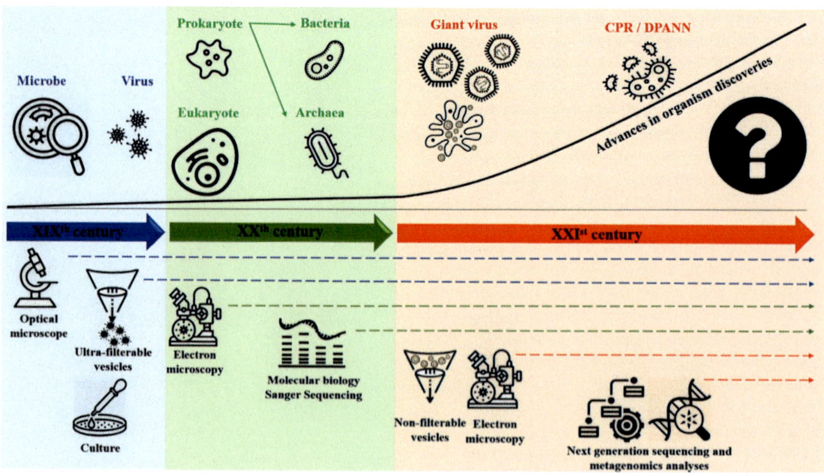

Figure 4.1. *Types of bacteria.*

Source: Ibrahim A, Colson P, Merhej V, Zgheib R, Maatouk M, Naud S, Bittar F, Raoult D. (2021). Rhizomal reclassification of living organisms. International Journal of Molecular Sciences, 22(11), 5643. https://doi.org/10.3390/ijms22115643.

. This diversity is critical for understanding the complexities of microbial life, ecosystem function, and potential bacterial applications in biotechnology, medicine, and environmental management

4.1.1. Taxonomic Diversity

Bacterial diversity refers to a wide range of species classified using various taxonomic methods. Traditional taxonomy, which is based on phenotypic characteristics such as morphology, staining properties, and metabolic capabilities, provided the foundation for understanding bacterial diversity. However, these methods are limited since they cannot resolve the vast diversity found in microbial communities. The introduction of molecular techniques, particularly 16S rRNA gene sequencing, transformed bacterial taxonomy by providing a solid framework for phylogenetic analysis (Vos et al., 2013). This gene, which is conserved across bacterial species, enables more accurate bacterial identification and classification. Studies using 16S rRNA sequencing have revealed an enormous diversity of uncultured bacterial lineages, highlighting the importance of comprehensive molecular approaches in microbial taxonomy.

4.1.2. Genetic Diversity

The ability of bacterial populations to adapt and evolve depends critically on their genetic diversity. Because of processes like recombination, mutation, and horizontal gene transfer (HGT), bacteria have a high degree of genetic variation. HGT, in particular, facilitates the exchange of genetic material between distantly related species, which is important for bacterial evolution (Vos et al., 2013). This process can confer resistance to antibiotics, new metabolic capacities, and other adaptive traits. The genome sizes of bacteria also exhibit genetic diversity; they can range greatly, from large, complex genomes in free-living species to small, streamlined genomes in obligate symbionts. Bacterial genomics research has shed light on the genetic foundation of bacterial diversity and its consequences for ecology and evolution (Murray et al., 2015).

4.1.3. Metabolic Diversity

Bacteria have remarkable metabolic diversity, allowing them to inhabit a variety of ecological niches. This diversity is evident in their ability to use a variety of energy sources and substrates, such as organic compounds, inorganic substances, and light. Photosynthetic bacteria, such as cyanobacteria, convert light energy into organic matter, playing an important role in global carbon cycling (Matulich et al., 2013). Chemolithotrophic bacteria derive energy from the oxidation of inorganic compounds, which helps to cycle elements like nitrogen, sulfur, and iron. Heterotrophic bacteria break down organic matter, recycling nutrients and promoting ecosystem health. This metabolic flexibility enables bacteria to adapt to a variety of environmental conditions and contribute to biogeochemical cycles, emphasizing their ecological importance (Madigan et al., 2018).

4.1.4. Ecological Roles

Bacteria play important ecological roles through their metabolic activities and interactions with other organisms. Bacteria in soil ecosystems decompose organic matter to promote nutrient recycling and soil formation (Torsvik and Øvreås, 2002). Nitrogen-fixing bacteria, such as rhizobia, form symbiotic relationships with leguminous plants and convert atmospheric nitrogen into ammonia, which plants can use. Bacteria in aquatic environments help to break down organic pollutants and cycle nutrients. The human microbiome, which is made up of various bacterial communities, is essential for health because it aids digestion, immune function, and pathogen protection (Vos et al., 2013). These ecological roles emphasize the importance of bacterial diversity in ensuring ecosystem stability and function.

4.1.5. Functional Diversity

Bacteria's functional diversity includes enzymatic activities, biochemical pathways, and ecological interactions. Bacteria produce a diverse range of enzymes that aid in the breakdown of complex organic compounds, the detoxification of harmful substances, and the synthesis of bioactive molecules (Vos et al., 2013). Biotechnology makes use of this functional versatility in applications such as bioremediation, in which bacteria degrade environmental pollutants, and industrial processes, in which bacterial enzymes catalyze chemical reactions. Metagenomics and metatranscriptomics, which analyze microbial populations' genetic and transcriptional profiles, are frequently used to assess the functional potential of bacterial communities. These methods reveal the functional capabilities and responses of bacterial communities to environmental changes (Madigan et al., 2018).

4.2. FUNGAL DIVERSITY

Fungal diversity is vast, encompassing a wide range of species classified into distinct groups.

Figure 4.2. *Fungal diversity.*

Source: Yadav N, Rakholia S, Ali N, Yosef R. (2023). Fungal diversity and distribution in the biodiversity hotspots of the Western Himalayas. Diversity, 15(11), 1106. https://doi.org/10.3390/d15111106.

4.2.1. Taxonomic Diversity

Traditionally, fungi were classified based on their morphological characteristics and reproductive structures. The main taxonomic groups are *Ascomycota, Basidiomycota, Zygomycota, Chytridiomycota*, and *Glomeromycota*. Molecular techniques, particularly DNA sequencing, have transformed fungal taxonomy by revealing cryptic species and improving phylogenetic relationships. Ribosomal RNA genes, such as the internal transcribed spacer (ITS) regions, have become the gold standard for fungal identification and classification (Maier et al., 2009). These molecular approaches have revealed a previously unknown diversity of fungi, greatly expanding our knowledge of fungal phylogeny and taxonomy.

4.2.2. Genetic Diversity

Genetic diversity within fungal populations is critical to their adaptability and evolutionary potential. Mutation, sexual and asexual reproduction, and HGT all contribute to fungi's significant genetic variation. Sexual reproduction, which involves the recombination of genetic material, produces genetic diversity and aids in the evolution of new traits. Furthermore, the presence of highly variable regions in fungal genomes, such as microsatellites and transposable elements, increases genetic diversity (van der Wal et al., 2013). Whole-genome sequencing and comparative genomics have shed light on the genetic complexity of fungi, revealing the existence of gene families involved in pathogenicity, symbiosis, and secondary metabolite production.

4.2.3. Metabolic Diversity

Fungi have remarkable metabolic diversity, which allows them to thrive in a variety of environments. They have a wide range of enzymatic capabilities that enable them to break down complex organic materials like lignin, cellulose, and chitin. This saprotrophic lifestyle is critical for nutrient cycling in ecosystems (Maier et al., 2009). Some fungi can also metabolize inorganic compounds like minerals, which contribute to biogeochemical processes. Furthermore, fungi participate in various types of symbiosis, such as mycorrhizal associations with plants, which improve nutrient uptake and plant growth. Fungi's metabolic versatility is demonstrated by their ability to produce a diverse range of secondary metabolites, such as antibiotics, toxins, and pigments, all of which have important ecological and biotechnological implications (Matulich et al., 2013).

4.2.4. Ecological Roles

Fungi play an important ecological role, contributing to ecosystem stability and function. Fungi act as decomposers, breaking down organic matter to aid in nutrient recycling and soil formation (van der Wal et al., 2013). Mycorrhizal fungi form symbiotic relationships with plant roots, which improve nutrient uptake and water absorption while also providing pathogen resistance. Endophytic fungi live in plant tissues and frequently provide benefits such as stress tolerance and increased growth. Pathogenic fungi, on the other hand, can cause diseases in plants, animals, and humans, having serious ecological and economic consequences (Maier et al., 2009). The ecological diversity of fungi emphasizes their critical role in maintaining ecosystem health and resilience.

4.2.5. Functional Diversity

Fungi are functionally diverse, with roles in nutrient cycling, symbiosis, pathogenesis, and secondary metabolite production. Fungal enzymes, such as cellulases, ligninases, and proteases, play an important role in the breakdown of complex organic compounds, helping to cycle carbon and nitrogen. Symbiotic fungi, such as mycorrhizal and endophytic species, help plants acquire nutrients and tolerate stress. Pathogenic fungi use a variety of mechanisms to infect host cells, evade immune responses, and cause disease. Furthermore, fungi produce a wide range of secondary metabolites with ecological functions, such as antimicrobial activity, and biotechnological applications, such as pharmaceuticals and industrial enzymes (Maier et al., 2009; van der Wal et al., 2013). The functional versatility of fungi emphasizes their ecological importance and biotechnological potential.

4.3. VIRAL DIVERSITY

Viral diversity refers to the wide range of genetic, structural, and functional variations found among viruses. Viruses, which are obligate intracellular parasites, differ greatly in their genetic makeup, morphology, replication strategies, and host range.

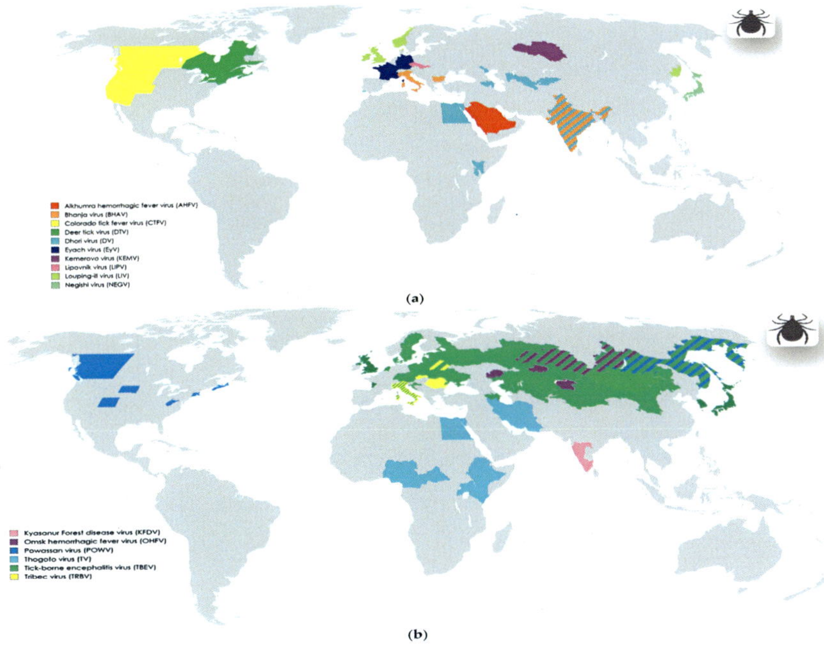

Figure 4.3. *Evolution has lead to a large diversity in the viral world.*

Source: Nurmukanova V, Matsvay A, Gordukova M, Shipulin G. (2024). Square the circle: Diversity of viral pathogens causing neuro-infectious diseases. Viruses, 16(5), 787. https://doi.org/10.3390/v16050787.

4.3.1. Taxonomic Diversity

Viruses have a wide taxonomic diversity, with families and genera based on morphology, nucleic acid type, replication mode, and host range. The International Committee on Taxonomy of Viruses (ICTV) is in charge of virus classification and naming, which is divided into orders, families, subfamilies, genera, and species (Lefkowitz et al., 2018). The major families are Herpesviridae, which contains large DNA viruses such as Herpes simplex virus (HSV), Retroviridae, which contains RNA viruses such as HIV, and Coronaviridae, which includes viruses, such as SARS-CoV-2. Advances in high-throughput sequencing and metagenomics have revealed a plethora of novel viruses, resulting in the identification of new families and challenging traditional taxonomic frameworks. This expanding taxonomic landscape highlights viral complexity and evolutionary diversity.

4.3.2. Genetic Diversity

Viral genetic diversity is remarkable, thanks to high mutation rates, genetic recombination, and reassortment. RNA viruses, in particular, evolve rapidly due to their error-prone RNA-dependent RNA polymerases (Murray et al., 2015). This genetic plasticity enables viruses to respond quickly to environmental changes and host defenses. For example, the influenza virus is subject to frequent genetic shifts and drifts, resulting in the emergence of new strains and the need for vaccine updates regularly. Furthermore, recombination and reassortment, as seen in coronaviruses and segmented RNA viruses such as rotaviruses, help to generate novel genotypes and phenotypes, increasing their adaptability and pathogenicity (Maier et al., 2009).

4.3.3. Metabolic Diversity

Viruses, unlike cellular organisms, lack their own metabolic machinery and rely solely on their host cells to replicate. However, they vary in how they exploit host cellular processes. DNA viruses, such as adenoviruses, frequently hijack the host's DNA replication machinery, whereas RNA viruses, such as poliovirus, replicate in the cytoplasm via viral RNA polymerases. Some viruses, such as retroviruses, incorporate their genetic material into the host genome, permanently changing the host cell's genetic landscape (Lefkowitz et al., 2018). Although viruses lack traditional metabolism, their ability to manipulate host metabolic pathways is diverse and sophisticated, demonstrating their adaptability to a wide range of hosts and cellular environments.

4.3.4. Ecological Roles

Viruses play an important role in ecological systems by influencing microbial diversity, nutrient cycling, and population dynamics. Bacteriophages, viruses that infect bacteria, are common in both marine and terrestrial environments and control bacterial populations through lysis (Maier et al., 2009). This viral predation regulates microbial community composition and increases nutrient turnover by lysing bacterial cells and releasing organic matter. Viruses also influence global biogeochemical cycles, particularly the carbon cycle, by altering the flow of carbon through microbial food webs. Furthermore, viruses can promote HGT within microbial communities, thereby increasing genetic diversity and evolution (Lefkowitz et al., 2018).

4.3.5. Functional Diversity

Viruses have functional diversity in terms of infection, immune evasion, and host interaction. Viruses have developed numerous strategies for entering host cells, replicating, and evading immune responses. For example, influenza viruses bind to host cell receptors using hemagglutinin proteins and release progeny virions via neuraminidase (Murray et al., 2015). HIV uses several mechanisms to evade the host immune system, including high mutation rates and the ability to remain latent in host cells. Viruses can also manipulate host cell processes to increase their replication and spread. Oncogenic viruses, such as human papillomavirus (HPV), can disrupt cell cycle regulation and cause cancer (Torsvik and Øvreås, 2002). This functional diversity emphasizes the complex interaction between viruses and their hosts, which drives viral evolution and pathogenesis.

4.4. PROTOZOAN DIVERSITY

Protozoan diversity refers to the variety of single-celled eukaryotic organisms found in the protozoa group. Protozoans are well-known for their intricate cellular structures, varied life cycles, and diverse ecological niches and host relationships.

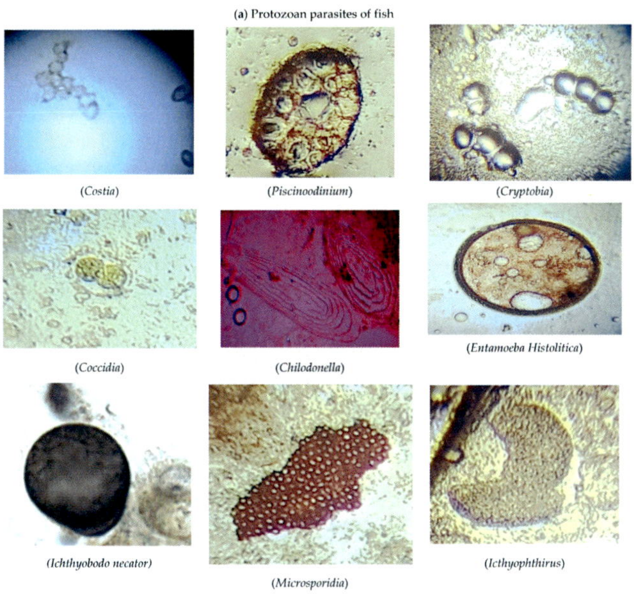

Figure 4.4. *Protozoan diversity.*

Source: Shafiq A, Abbas F, Hafeez-ur-Rehman M, Khan BN, Aihetasham A, Amin I, Hmidullah, Mothana RA, Alharbi MS, Khan I, et al. (2023). Parasite diversity in a freshwater ecosystem. Microorganisms, 11(8):1940. https://doi.org/10.3390/microorganisms11081940.

This diversity is important for understanding their roles in ecosystems, disease pathogenesis, and evolutionary biology.

4.4.1. Taxonomic Diversity

Protozoa are a group of unicellular eukaryotes classified according to their morphology, locomotion, and life cycles. Amoebozoa, which move with pseudopodia; Ciliophora, which have cilia; Apicomplexa, which have parasitic lifestyles and complex life cycles; and Euglenozoa, which have flagella (Maier et al., 2009). Advances in molecular phylogenetics have reshaped protozoan taxonomy, revealing deep evolutionary relationships and cryptic biodiversity. Ribosomal RNA (rRNA) gene sequencing has helped to identify new lineages and reclassify groups based on genetic data rather than morphological characteristics alone. This molecular approach has resulted in a more accurate and complete understanding of protozoan diversity.

4.4.2. Genetic Diversity

Protozoa have a high level of genetic diversity, which reflects their long evolutionary history and adaptability to a variety of environments. Mutation, genetic recombination, and gene flow all contribute to significant genetic variation within and between species of protozoa. Plasmodium, the parasite that causes malaria, for example, has a large genetic diversity, which contributes to its adaptability and drug resistance (Manske et al., 2012). Whole-genome sequencing and comparative genomics have shed light on the genetic basis of protozoan diversity, revealing gene families involved in host interaction, immune evasion, and metabolic adaptation (Maier et al., 2009). This genetic complexity demonstrates protozoa's evolutionary success and ecological versatility.

4.4.3. Metabolic Diversity

Protozoa have a wide range of metabolic capabilities, allowing them to live in a variety of ecological niches. They may be autotrophic, heterotrophic, or mixotrophic. Autotrophic protozoa, such as Euglenoids, use chloroplasts for photosynthesis (Maier et al., 2009). Heterotrophic protozoa, such as amoebae and ciliates, consume bacteria, algae, and other small particles via phagocytosis. Mixotrophic protozoa combine photosynthesis and heterotrophy, enabling them to adapt to changing environmental conditions. Furthermore, some protozoa have specialized metabolic pathways to survive in extreme environments, such as anaerobic conditions, where they engage in anaerobic respiration or fermentation (Murray et al., 2015).

4.4.4. Ecological Roles

Protozoa play an important role in ecosystems, particularly nutrient cycling and food web dynamics. As primary consumers, they regulate bacterial populations and recycle nutrients through predatory behavior. Protozoa in aquatic ecosystems help to close the microbial loop, allowing organic matter and energy to move from bacteria to higher trophic levels. Soil protozoa improve soil fertility by breaking down organic matter and stimulating microbial activity. Parasitic protozoa from the phylum Apicomplexa have a significant impact on host populations and ecosystem health, causing diseases in humans, animals, and plants (Manske et al., 2012; Matulich et al., 2013). Protozoa play a variety of ecological roles that are critical to the functioning and stability of ecosystems.

4.4.5. Functional Diversity

Protozoa are functionally diverse, with roles in symbiosis, pathogenesis, and environmental adaptation. Symbiotic protozoa, such as *Trichonympha* in termite guts, aid in the digestion of cellulose, allowing termites to feed on wood (Maier et al., 2009). Pathogenic protozoa, such as *Trypanosoma* and *Giardia*, have complex life cycles and evasion strategies for infecting hosts and avoiding immune responses. Acanthamoeba, an environmental protozoan, forms cysts to protect itself from harsh conditions such as desiccation and temperature extremes. The functional versatility enables protozoans to thrive in a variety of environments and interact with other organisms in a variety of ways.

4.5. ALGAL DIVERSITY

Algae are a diverse group of photosynthetic organisms found in both aquatic and terrestrial environments.

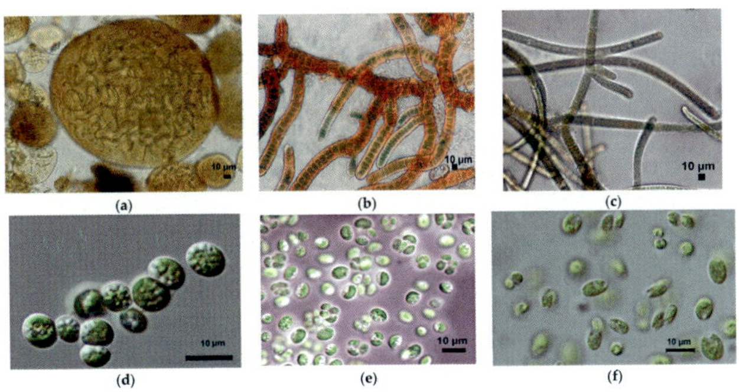

Figure 4.5. *Algal diversity.*

Source: Patova E, Novakovskaya I, Gusev E, Martynenko N. (2023). Diversity of cyanobacteria and algae in biological soil crusts of the northern ural mountain region assessed through morphological and metabarcoding approaches. Diversity, 15(10):1080. https://doi.org/10.3390/d15101080.

They range from microscopic single-celled organisms to large multicellular forms, with a wide range of morphology, pigmentation, life cycles, and ecological roles. This diversity is critical for understanding their effects on ecosystems, biogeochemical cycles, and potential biotechnological applications.

4.5.1. Taxonomic Diversity

Pigmentation, storage products, cell wall composition, and other morphological and genetic characteristics distinguish algae into several major groups. The primary groups are *Chlorophyta* (green algae), *Rhodophyta* (red algae), *Phaeophyceae* (brown algae), and *Bacillariophyta* (diatoms), among others. Advances in molecular phylogenetics have significantly refined algal taxonomy, revealing evolutionary relationships and reclassifying many species. Ribosomal RNA (rRNA) and plastid gene sequencing have been especially useful in resolving algal phylogeny, revealing cryptic diversity, and providing insights into evolutionary history (Maier et al., 2009).

4.5.2. Genetic Diversity

Algae have a high degree of genetic diversity, which reflects their long evolutionary history and adaptation to various habitats. The size and complexity of algal genomes vary greatly, as do their gene content and organization. Whole-genome sequencing projects have revealed the genetic basis for important traits like photosynthesis, nutrient acquisition, and stress responses. For example, the genomes of model green algae like *Chlamydomonas reinhardtii* and red algae like *Cyanidioschyzon merolae* have shed light on the evolution of photosynthetic machinery and metabolic pathways. Furthermore, HGT from bacteria and other organisms has increased algal genetic diversity, improving their adaptability (Matulich et al., 2013).

4.5.3. Metabolic Diversity

Algae have remarkable metabolic diversity, allowing them to inhabit a variety of ecological niches. They are primary producers, capturing light energy through oxygenic photosynthesis with chlorophyll and accessory pigments. Algae have a variety of metabolic pathways in addition to photosynthesis, including carbon fixation, nitrogen assimilation, and lipid synthesis. Some algae, such as diatoms and dinoflagellates, have distinct metabolic capabilities, including the production of biogenic silica and bioluminescence (Maier et al., 2009). Furthermore, many algae can switch between autotrophic and heterotrophic modes of nutrition in response to environmental conditions, a phenomenon known as mixotrophy. This metabolic flexibility contributes to their environmental success and resilience.

4.5.4. Ecological Roles

Algae play important ecological roles in both aquatic and terrestrial ecosystems. As primary producers, they serve as the foundation of the food web, sustaining a variety of marine and freshwater communities. Phytoplankton, which includes diatoms, dinoflagellates, and cyanobacteria, play important roles in marine primary production, driving global carbon and nutrient cycles. Algae produce oxygen through photosynthesis, which has a significant impact on atmospheric composition. In addition to their roles in aquatic systems, algae form symbiotic relationships with other organisms, such as lichens (fungi and green algae or cyanobacteria) and corals (zooxanthellae, a type of dinoflagellate) (Maier et al., 2009). These interactions are critical to the health and stability of ecosystems.

4.5.5. Functional Diversity

Algae's functional diversity includes roles in nutrient cycling, symbiosis, and biotechnology applications. Algae produce a diverse range of bioactive compounds, including vitamins, antioxidants, and polysaccharides, which are both ecologically and economically important. Their ability to sequester carbon and produce biofuels and bioplastics has sparked interest in their potential to combat climate change and provide renewable energy sources. Algae are versatile and useful in a variety of applications, including wastewater treatment and environmental health monitoring (Maier et al., 2009; Torsvik and Øvreås, 2002). Understanding and harnessing algae's functional diversity is critical for addressing environmental issues and advancing biotechnological innovations.

4.6. ARCHAEA DIVERSITY

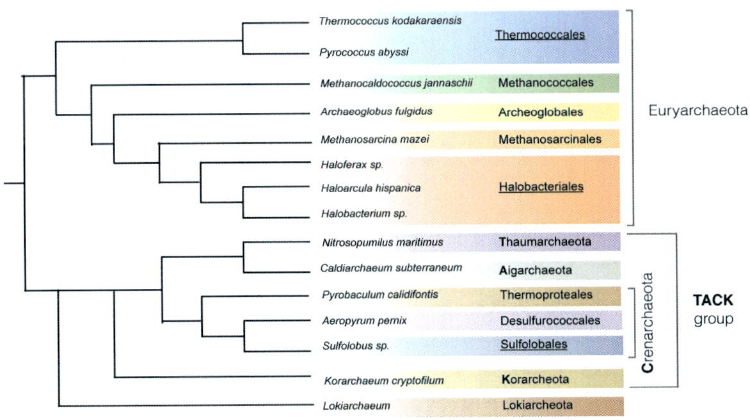

Figure 4.6. *Archaeal diversity and their current phylogenetic relationship.*

Source: Ausiannikava D, Allers T. (2017). Diversity of DNA replication in the archaea. Genes, 8(2):56. https://doi.org/10.3390/genes8020056.

Archaeal diversity encompasses a wide range of taxa, genetic variations, metabolic pathways, ecological niches, and functional roles. This diversity highlights the adaptability and evolutionary significance of Archaea in various environments.

4.6.1. Taxonomic Diversity

Among the three domains of life, along with bacteria and Eukarya, Archaea has an impressively diverse taxonomic range. They belong to a number of significant phyla, such as *Asgardarchaeota*, which was only recently identified, and *Euryarchaeota*, *Crenarchaeota*, *Thaumarchaeota*, and *Nanoarchaeota*. Many genera and species that have adapted to different environmental niches make up each phylum. Methanogens, for example, are organisms that make methane, and halophiles are organisms that are found in habitats with high salinity. Thermophiles and hyperthermophiles that live in extremely hot places like hot springs and hydrothermal vents make up the majority of the *Crenarchaeota*. Given that *Asgardarchaeota* and eukaryotes have genetic similarities, this discovery has shed light on the evolution of complex life (Spang et al., 2015).

4.6.2. Genetic Diversity

Archaea's genetic diversity is tremendous, demonstrating their adaptability to a variety of settings. Comparative genomics has shown considerable differences in genome size, gene content, and regulatory mechanisms between archaeal species. HGT has played an important role in modifying archaeal genomes, allowing for the development of new metabolic capacities and adaptations. Genes involved in methanogenesis, a distinct metabolic process seen in some archaea, are thought to have been acquired by HGT (Torsvik and Øvreås, 2002). Furthermore, the presence of unique genetic elements, such as CRISPR-Cas systems in archaea underlines their adaptive immunity against viruses and plasmids, which contributes to their genetic diversity (Vos et al., 2013).

4.6.3. Metabolic Diversity

Archaea have a complex set of metabolic pathways that allow them to flourish in a variety of settings, including many harsh ones. They are recognized for their ability to use a wide range of energy sources and metabolic processes, such as methanogenesis, sulfur reduction, and ammonia oxidation. Methanogens, primarily found in the *Euryarchaeota* phylum, create methane as a byproduct of anaerobic respiration and play an important part in the global carbon cycle. Sulfur-reducing archaea, such as those from the genus *Archaeoglobus*, use sulfur compounds as electron acceptors in their energy metabolism. *Thaumarchaeota*, a phylum of AOA, contribute to the nitrogen cycle by converting ammonia to nitrite (Spang et al., 2015). This metabolic plasticity highlights the ecological significance of archaea in biogeochemical cycles.

4.6.4. Ecological Roles

Archaea live in a variety of ecological niches, ranging from harsh conditions to more moderate ones. They can be found in hot springs, acidic mines, hypersaline lakes, deep-sea hydrothermal vents, soil, and even the ocean. This capacity to live in such various habitats is due to its distinct physiological and metabolic adaptations. Halophilic archaea, such as those found in the Halobacteria class, have systems in place to maintain osmotic balance in high-salt environments, such as the manufacture of suitable solutes and specific membrane proteins (Vos et al., 2013). In marine habitats, archaeal communities, particularly those of *Thaumarchaeota*, play an important role in nutrient cycling and primary production (Torsvik and Øvreås, 2002).

4.6.5. Functional Diversity

Archaea play an important role in many ecological processes and biogeochemical cycles. Their roles in methanogenesis, nitrogen cycling, and sulfur reduction are well understood and have important environmental ramifications. For example, methanogenic archaea play an important role in the decomposition of organic matter in anaerobic environments, contributing to methane emissions, a strong greenhouse gas (Vos et al., 2013). *Nitrosopumilus maritimus* and other AOA play important roles in the nitrogen cycle, regulating nitrogen availability and ecosystem production. Furthermore, the revelation of archaeal involvement in hydrocarbon breakdown and bioremediation indicates their potential uses in environmental biotechnology (Spang et al., 2015).

4.7. EXTREMOPHILES

Extremophiles are organisms that can be found in extreme environmental conditions. The name of this group was derived from the Greek word 'philos' which means 'lover' of extreme conditions. These organisms exhibit unique adaptations that enable survival where most life forms cannot persist (Udaondo et al., 2016).

Figure 4.7. *Diversity of extreme environments on Earth.*

Source: Vavitsas K, Glekas PD, Hatzinikolaou DG. (2022). Synthetic biology of thermophiles: Taking bioengineering to the extremes? Applied Microbiology, 2(1), 165-174. https://doi.org/10.3390/applmicrobiol2010011.

4.7.1. Types of Extremophiles

Extremophiles are classed based on the conditions in which they grow. Typically, ecosystems have a variety of physiochemical conditions, necessitating extremophile adaptation to numerous physiochemical parameters. Extremophiles encountered in such settings are known as "polyextremophiles."

1. **Acidophiles**: Acidophiles are accustomed to environments with acidic pH levels ranging from 1 to 5. This group contains eukaryotes, bacteria, and archaea found in sulfuric pools, acidic mine drainage sites, and even our stomachs (Cavicchioli, 2016).

Acidophiles maintain their pH levels via several specialized methods, some passive (not exerting energy) and some active (exerting energy). Passive techniques often include strengthening the cell barrier against the external environment, which may include secreting a biofilm to prevent chemicals from diffusing into the cell or completely altering the cell membrane to incorporate protective compounds like fatty acids. Some acidophiles can produce buffer molecules to assist elevate their internal pH. Active pH regulation systems use a hydrogen ion pump to remove hydrogen ions from the cell at a continuously high rate (Udaondo et al., 2016).

2. **Alkaliphiles**: Alkaliphiles thrive in environments with basic pH levels of 9 or above. They employ both passive and active methods to preserve homeostasis. Cell internal pooling of cytoplasmic polyamines is one example of a passive process. Because of their abundance in positively charged amino groups, polyamines act as a buffer for the cytoplasm in alkaline conditions. A low membrane permeability is another passive mechanism that prevents protons from entering and leaving the cell. Protons enter cells through a sodium ion channel, which is the active mechanism of control (Cavicchioli, 2016).
3. **Thermophile**: Thermophiles are organisms that can withstand temperatures as high as 251°F. Hot springs, volcanic deposits, and hydrothermal vents are a few locations where they can be found. Their extremozymes are responsible for their survival in these environments. These kinds of enzymes can continue to work properly even under extremely hot conditions because their amino acids do not misfold or lose their form (Lederberg and Relman, 2009).
4. **Psychrophile**: Psychrophiles, sometimes referred to as cryophiles, are able to survive in extremely low temperatures, down to 5°F. This group can be found in cold soils, permafrost, polar ice, cold ocean water, and alpine snow packs. It is a part of all three domains of life (bacteria, archaea, and Eukarya).

Their extremozymes, which continue to function at low temperatures and a little more slowly at even lower temperatures, are one reason they can withstand extremely cold temperatures. Additionally, the proteins produced by psychrophiles can operate in cold environments and have high levels of unsaturated fatty acids in their plasma membranes, which act as a buffer for the cells against the cold. Most remarkably, though, some psychrophiles can substitute the sugar trehalose for the water in their bodies, which stops hazardous ice crystals from forming (Udaondo et al., 2016).

5. **Xerophiles**: The exceedingly dry environments that can be exceedingly hot or cold are home to xerophiles. They have been discovered in the Antarctic, the Great Basin, and the Atacama Desert. Certain xerophiles, like their psychrophilic counterparts, can substitute trehalose for water, which can shield membranes and other structures from times when there isn't as much water available.
6. **Barophile (Piezophile):** Organisms that thrive at pressures of 400 atm or higher are known as barophiles. They can live by controlling the membrane's phospholipids' fluidity. The external environment and the pressure gradient between the inside and exterior of the cell are balanced by this fluidity. Extreme barophiles cannot grow at lower pressures; they grow ideally at 700 atm or more (Cavicchioli, 2016).
7. **Halophile**: Halophiles are organisms that grow best in environments with high salt concentrations. Most common are prokaryotic bacteria at salinities higher than 1.5M. Though in fewer numbers, this group nevertheless encompasses all three spheres of existence. The initial step in overcoming the challenges of hypersaline conditions is to reduce the amount of water lost by cells. Halophiles accomplish this by using a variety of methods to build up solutes in the cytoplasm. A sodium-potassium ion pump is used by halophilic archaea to take in potassium and eliminate sodium. Glycerol is used by halotolerant bacteria as a suitable solute to maintain osmotic pressure balance (Lederberg and Relman, 2009).

4.7.2. Importance of Extremophiles

Extremophiles are important in microbiology because they reveal the adaptability of life and have potential applications in biotechnology. These microbes flourish in severe environments, offering novel enzymes and bioactive chemicals for commercial application and expanding our understanding of life's resilience, which is vital for astrobiology and evolutionary studies (Cavicchioli, 2016).

4.8. SYMBIOTIC RELATIONSHIPS

A symbiotic relationship is an intimate relationship between two or more species in which at least one benefits. These interactions are critical to the survival and function of many ecosystems. Symbiotic connections in microbiology frequently involve microorganisms including bacteria, fungi, and viruses interacting with plants, animals, and other microbes. These interactions can be either obligatory, in which the species cannot exist without one another, or facultative, in which the organisms can survive independently but benefit from the affiliation (Cordell, 2000).

4.8.1. Types of Symbiotic Relationships

4.8.1.1. Mutualism

Mutualism is a kind of symbiotic interaction in which each of the species involved get something from the partnership. The survival and growth of the organisms frequently depend on this interaction. The interaction between leguminous plants and Rhizobium bacteria is a prime example of mutualism. Leguminous plants have root nodules where rhizobium bacteria live and fix atmospheric nitrogen into a form the plant can absorb and use for growth. The plant provides the bacteria with a protective environment and sugars in exchange. For agricultural production and soil fertility, particularly in legume crops, this mutualistic relationship is essential (Guan and Liu, 2020).

The association between some fungi and algae or cyanobacteria in lichens is another illustration of mutualism. While the algae or cyanobacteria carry out photosynthesis to produce carbohydrates that feed the fungi, the fungi provide a suitable home and absorb nutrients. Lichens are able to colonize and flourish in a variety of typically hostile habitats because of their mutualistic connection (Pérez-Brocal et al., 2006).

4.8.1.2. Commensalism

Commensalism is the term used to characterize a connection in which one organism gains while the other experiences neither damage nor benefit. The coexistence of the human host and the bacteria *Staphylococcus epidermidis* is an illustration of commensalism. The human host is not harmed, and the bacteria gain from the skin's provision of nutrients and a suitable environment. In certain cases, these beneficial bacteria can even provide defense by outcompeting hazardous diseases for resources and available space. It is common to find commensal partnerships between different microbes and larger hosts in maritime environments. As an example, certain bacteria adhere to the surface of sea algae or animals and take advantage of the nutrients and surface area that the host provides, all without harming the host (Cordell, 2000).

4.8.1.3. Parasitism

In a symbiotic connection, one organism—the parasite—benefits at the expense of the other—the host. This is known as parasitism. Relationships with parasites can cause the host to suffer from modest discomfort to serious illness or even death. The bacteria that causes tuberculosis, Mycobacterium tuberculosis, is a prime example of parasitism. The bacteria takes up residency in the host tissue and produces a serious respiratory infection when it infects human lungs.

The protozoan parasites *Plasmodium* spp., which cause malaria in humans, are a further illustration. Through the bites of infected Anopheles mosquitoes, these parasites are spread. Plasmodium species grow and infiltrate red blood cells within the human host, resulting in symptoms including fever, chills, and anemia that have a serious negative effect on the host's health (Pérez-Brocal et al., 2006).

4.8.2. Importance of Symbiotic Relationships

The maintenance of biodiversity and the operation of ecosystems depend on symbiotic relationships. They improve soil fertility, aid in the cycling of nutrients, and promote the survival and well-being of a variety of creatures (Pérez-Brocal et al., 2006). Both parties in mutualistic partnerships frequently develop adaptations that maximize their shared advantages, which increases specialization and reliance.

Understanding and utilizing mutualistic interactions in agriculture, such as the one between legumes and bacteria that fix nitrogen, can increase crop yields and decrease the need for artificial fertilizers. The study of commensal microorganisms in human health can result in the creation of probiotics and other treatments that increase populations of helpful bacteria while averting or treating diseases brought on by harmful microbes.

Although parasitic relationships are frequently harmful to the host, they are the driving force for the development of host defense mechanisms and parasite tactics to get past these barriers. A deeper comprehension of disease processes as well as the creation of novel treatments and preventative measures may result from this dynamic interaction (Pérez-Brocal et al., 2006).

4.9. MICROBIAL EVOLUTION

In microbiology, microbial evolution is an important field of study that explores the genetic and phenotypic changes that bacteria undergo over time. The mechanisms behind evolutionary processes, including genetic drift, natural selection, mutation, and HGT, are covered in this topic. Comprehending the evolution of microbes is essential to understanding the emergence of antibiotic resistance, disease adaptation, and ecological dynamics. It also sheds light on the beginnings of life, evolutionary biology, and the uses of biotechnology (Lederberg and Relman, 2009). By clarifying these evolutionary routes, scientists can create novel approaches to industrial microbiology, the fight against infectious diseases, and the use of microbes for improvements in the environment and medicine.

4.9.1. Mechanisms of Evolution

The various processes that propel genetic diversity and adaptation in microbes are referred to as microbial evolution. Natural selection, genetic recombination, HGT, and mutation are important processes.

- **Mutation**: In microbial communities, spontaneous mutations are the main source of genetic variation. Errors in DNA replication or exposure to mutagenic chemicals can result in mutations, which generate novel alleles that may be vulnerable to selection pressures. Long-term evolution experiments with *Escherichia coli* have shown advantageous mutations accumulate over thousands of generations, boosting fitness and adaptability (Vos, 2009).
- **Horizontal Gene Transfer**: One essential method by which bacteria obtain genetic material from unrelated species is called horizontal gene transfer, or HGT. Conjugation, transduction, or transformation are the processes that occur in this evolution mechanism. The important significance of HGT in bacterial evolution bringing new metabolic pathways and antibiotic resistance genes hence HGT promotes quick adaption (Lederberg and Relman, 2009).
- **Genetic Recombination**: This process of exchanging DNA segments between distinct genomes results in new genetic combinations. It is most commonly observed in bacteria. Either HGT or spontaneous recombination of homologous sequences can trigger the process. Recombination rates differ throughout bacterial species, affecting their ecological niches and evolutionary paths (Hewitt, 2004).
- **Natural Selection**: influences genetic variation within microbial populations, favoring characteristics that improve survival and reproduction in specific conditions. For example, the evolution of antibiotic resistance in bacteria is an obvious example of natural selection in action. Lederberg and Relman (2009) explain how the selective forces exerted by antibiotic use in clinical and agricultural settings have influenced the emergence of resistant organisms. These systems allow microbes to adapt quickly to changing surroundings, contributing to their evolutionary success and ecological diversity.

4.9.2. Microbial Phylogenetics and Evolutionary Relationships

By examining the evolutionary connections between microorganisms, microbial phylogenetics sheds light on their genetic diversity and evolutionary background. Phylogenetic trees that clarify the lineage-specific adaptations and HGT events that influence microbial evolution can be created by researchers through the analysis of genetic sequences. Hug et al. (2016) conducted a study that emphasizes the vast genetic variety and intricate evolutionary processes seen in the microbial realm. The authors created a tree of life using a complete set of genomes, exposing previously unknown branches and indicating that a large portion of microbial diversity is yet unknown. This research underscores the significance of phylogenetic analysis in understanding microbial evolution, as it allows for the identification of evolutionary trends and the elucidation of the mechanisms driving microbial diversification. Such knowledge is crucial for fields ranging from ecology to medicine, as it aids in the development of targeted therapies and the prediction of microbial responses to environmental changes (Lederberg and Relman, 2009).

4.9.3. Application and Implication of Microbial Evolution

Microbial evolution has broad applications and ramifications that impact domains like environmental science, agriculture, and health. Natural selection and genetic variation promote microbial evolution, which allows bacteria to adapt to a variety of settings and give rise to novel dangerous strains and antibiotic resistance. It is essential to comprehend and control these evolutionary dynamics to treat infectious diseases. For example, a study by Madigan et al. (2018) highlights how gene exchange among bacterial populations accelerates the transmission of antibiotic resistance genes and clarifies the function of HGT in bacterial adaptation. This has significant effects on public health and calls for the creation of new surveillance and antimicrobial policies. Biotechnological applications, such as the engineering of microorganisms for industrial processes and bioremediation, are also informed by the evolution of bacteria. Thus, addressing the benefits and difficulties brought about by microbial diversity and adaptation requires an understanding of microbial evolution (Madigan et al., 2018).

4.10. MICROBIAL BIOTECHNOLOGY

Microbial biotechnology uses microorganisms' capabilities to create processes and products with far-reaching ramifications in a wide range of industries. This dynamic and diverse subject applies microbiology, genetics, and biochemical engineering principles to improve industrial operations, environmental sustainability, agricultural productivity, and medicinal advances. In this discussion, we will look at the various subfields of industrial biotechnology, environmental biotechnology, agricultural biotechnology, and medical biotechnology, focusing on the roles and advancements that microbial biotechnology has driven in each industry.

4.10.1. Industrial Biotechnology

Industrial biotechnology (also known as white biotechnology) uses microorganisms to produce chemicals, biofuels, and enzymes, resulting in more sustainable industrial operations. Microorganisms such as bacteria, yeast, and fungi are used to turn basic materials into valuable products via fermentation and other biotechnological processes. For example, the generation of bioethanol from agricultural waste using modified strains of *Saccharomyces cerevisiae* is an industrial use of microbial biotechnology (Zhang et al., 2018).

Enzymes generated from microorganisms serve an important role in many industrial processes. Amylases, proteases, and lipases are commonly employed in the food business to perform starch hydrolysis, protein degradation, and fat modification. These microbial enzymes not only increase process efficiency but also reduce environmental impact by eliminating the need for harsh chemicals and energy-intensive methods.

4.10.2. Environmental Biotechnology

Environmental biotechnology uses the metabolic variety of microorganisms to address environmental issues such as pollution, waste management, and resource recovery. Bioremediation, a critical component of environmental biotechnology, is the employment of microbes to degrade or detoxify contaminants in soil, water, and the air. Certain bacteria, for example, can breakdown hydrocarbons in oil spills, reducing the toxicity of dangerous compounds (Das and Chandran, 2011).

Furthermore, microbial biotechnology facilitates wastewater treatment by breaking down organic materials and removing pollutants. Microbial fuel cells (MFCs) are a novel technology in which microorganisms produce electricity while treating wastewater, so combining waste management and energy generation (Logan et al., 2006). This dual functioning demonstrates the potential of microbial processes to promote environmental sustainability.

4.10.3. Agricultural Biotechnology

Microorganisms in agricultural biotechnology improve crop output, soil health, and insect management, all of which contribute to more sustainable farming methods. Plant growth-promoting rhizobacteria (PGPR) are helpful microbes that colonize plant roots and promote development by generating phytohormones, fixing atmospheric nitrogen, and solubilizing vital nutrients (Lugtenberg and Kamilova, 2009). These microbial interactions not only increase crop yields but also minimize the need for chemical fertilizers. Microbial biotechnology also addresses plant diseases with biocontrol agents. Fungi and bacteria can combat plant pathogens, lowering disease rates and reducing the demand for chemical pesticides. For example, *Bacillus subtilis* produces antibiotics that reduce fungal infections, providing an environmentally acceptable alternative to traditional agrochemicals.

4.10.4. Medical Biotechnology

Medical biotechnology uses microorganisms to generate diagnostics, therapies, and vaccinations, changing healthcare. Antibiotic synthesis is one of the earliest and most significant achievements of microbial biotechnology. Penicillin, produced from the fungus *Penicillium notatum*, heralded the antibiotic age by significantly lowering mortality from bacterial infections (Fleming, 1929).

Modern breakthroughs include the use of genetically modified microbes to manufacture insulin, a critical hormone for diabetics. Recombinant DNA technology allows for the introduction of human insulin genes into *Escherichia coli*, enabling large-scale synthesis of insulin of high purity and efficacy. Furthermore, microbial biotechnology plays a pivotal role in vaccine development. Attenuated or inactivated pathogens and recombinant vaccines rely on microbial systems to elicit immune responses, protecting against infectious diseases.

CHAPTER 5

MICROBIAL ECOLOGY

Contents

Microbial ecology is a microbiology concept that studies the complex interactions between microorganisms and their environments, which include both biotic and abiotic components. This discipline investigates the roles of microbes in nutrient cycling, ecosystem function, and biogeochemical processes. Vos et al. (2013) investigated how microbial communities influence and are influenced by their habitats, resulting in evolutionary adaptations and ecological dynamics. The combination of high-throughput sequencing technologies and bioinformatics tools has transformed our knowledge of microbial diversity, community structure, and functional capabilities (Anderson, 2003). Furthermore, microbial ecology's understanding of symbiotic relationships, pathogenicity, and microbial resilience highlights its importance in addressing global issues such as climate change, public health, and sustainable agriculture. Thus, microbial ecology not only broadens our understanding of microbial life, but also promotes progress in biotechnology, environmental management, and medical sciences (Maier et al., 2009).

5.1. MICROBIAL HABITATS

Microorganisms inhabit diverse environments, each uniquely adapted to their specific ecological niches. Their habitats are remarkably varied, reflecting their physiological and metabolic versatility.

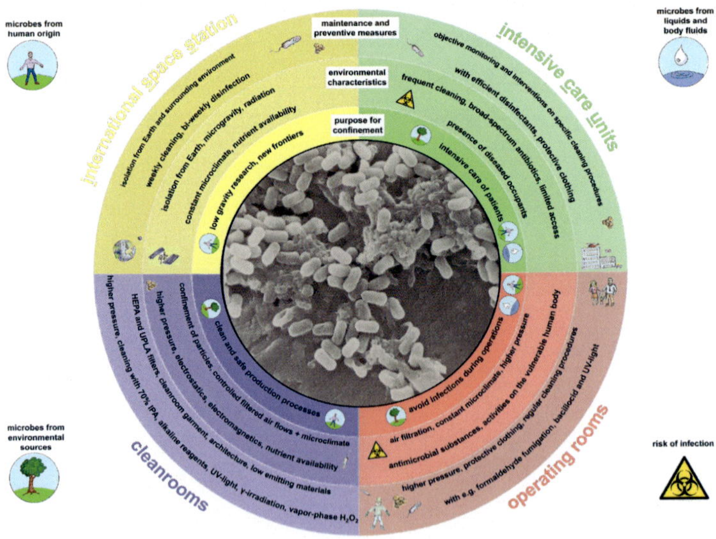

Figure 5.1. *The diverse habitats in which microbes live and exert important influence.*

Source: Mora M, Mahnert A, Koskinen K, Pausan MR, Oberauner -Wappis L, Krause R, Perras AK, Gorkiewicz G, Berg G and Moissl -Eichinger C (2016) Mi-

croorganisms in confined habitats: Microbial monitoring and control of intensive care units, operating rooms, cleanrooms and the international space station. Front. Microbiol., 7,1573. doi: 10.3389/fmicb.2016.01573.

5.1.1. Bacteria

Bacteria are ubiquitous, thriving in almost every environment on Earth. They colonize soil, water, air, and the surfaces of living and non-living things. Bacteria such as Nitrosomonas and Nitrobacter are important nitrogen cycling agents in soil. Aquatic environments, from freshwater to the deep sea, support a wide range of bacterial communities that recycle nutrients and decompose organic matter. Extremophiles, like Thermus aquaticus, thrive in hot springs and hydrothermal vents, demonstrating bacterial adaptability to harsh environments (Hurst et al., 2007; Maier et al., 2009; Lievens et al., 2015). Furthermore, human and animal microbiota, which include gut bacteria such as *Escherichia coli* and Bacteroides, are critical for digestion, immunity, and overall health.

5.1.2. Fungi

Fungi are primarily found in terrestrial environments, where they act as decomposers, symbionts, and pathogens. Soil and leaf litter support a diverse range of fungal species, including Penicillium and Aspergillus, which decompose organic matter and recycle nutrients. Mycorrhizal fungi form symbiotic relationships with plant roots to improve water and nutrient uptake. Fungi also live in decaying wood and compost, where species such as *Trametes versicolor* degrade lignin and cellulose. Fungi such as Chytridiomycetes decompose organic material in both freshwater and marine ecosystems. Pathogenic fungi, such as *Candida albicans* and *Cryptococcus neoformans*, can cause illness in both humans and animals.

5.1.3. Viruses

Viruses are obligate intracellular parasites that can infect all forms of life, including bacteria and humans. Bacteriophages, or viruses that infect bacteria, are abundant in soil and water, influencing microbial population dynamics and gene transfer. Human viruses, such as influenza and HIV, replicate in a variety of tissues and organs within the host using cellular machinery (Maier et al., 2009). Plant viruses, such as tobacco mosaic virus, can be found in plant tissues and are frequently transmitted by insects. Marine viruses play an important role in controlling phytoplankton populations and nutrient cycling in oceanic ecosystems (Matulich et al., 2013).

5.1.4. Protozoa

Protozoa are primarily aquatic microorganisms that live in freshwater and marine environments, as well as moist soil. Free-living protozoa, such as Paramecium and Amoeba, live in ponds, streams, and oceans, where they consume bacteria, algae, and detritus. Some protozoa, such as Plasmodium and Trypanosoma, are parasitic, living in vertebrate hosts' blood and tissues and causing diseases like malaria and sleeping sickness (Vos et al., 2013). Soil protozoa play an important role in nutrient cycling by feeding on bacteria and other microorganisms.

5.1.5. Algae

Algae are primarily photosynthetic organisms that live in aquatic environments such as freshwater lakes, rivers, and the ocean. Marine algae, such as diatoms and dinoflagellates, are essential components of phytoplankton, serving as the foundation of the marine food web and contributing significantly to global primary production. Freshwater algae, such as green algae and cyanobacteria, live in ponds, streams, and lakes, and they frequently form blooms in nutrient-rich environments. Terrestrial algae form lichens on soil, rocks, and tree trunks through symbiotic relationships with fungi (Torsvik and Øvreås, 2002; Maier et al., 2009).

5.1.6. Archaea

Archaea are extremophiles that live in some of the harshest environments on Earth. Thermophilic archaea, like Sulfolobus and Pyrococcus, thrive in hot springs and hydrothermal vents with temperatures above 100°C (Maier et al., 2009). Halophilic archaea, such as Halobacterium, thrive in highly saline environments like salt lakes and solar salterns. Methanogenic archaea that produce methane live in anaerobic environments such as wetlands, ruminant digestive tracts, and anaerobic sludge digesters. Archaea can also be found in more moderate environments, such as soils and oceans, where they contribute to nutrient cycling and biogeochemical processes (Hurst et al., 2007).

5.2. MICROBIAL INTERACTIONS

Microbial interactions are intricate and multifaceted, influencing ecological balance and biogeochemical cycles. These interactions, which include bacteria, fungi, viruses, protozoa, algae, and archaea, can be synergistic, antagonistic, or neutral, influencing the dynamics of microbial communities and their environments (Hurst et al., 2007).

5.2.1. Bacteria-Fungi Interactions

Bacteria and fungi frequently coexist in soil and plant environments, and their interactions can be both competitive and cooperative. Mycorrhizal fungi and PGPR have mutualistic relationships in which fungi improve nutrient uptake for plants and bacteria produce growth hormones and suppress pathogens. In resource competition, however, bacteria and fungi compete for organic substrates, resulting in antagonistic interactions. For example, Pseudomonas species produce antifungal compounds that inhibit fungal pathogens and thus protect plants from disease (Amin et al., 2015).

5.2.2. Bacteria-Virus Interactions

Viruses, particularly bacteriophages, are important regulators of bacterial populations. Phages infect and kill bacteria, affecting microbial community structure and nutrient cycling. This interaction is an important part of the microbial loop, as viral lysis releases organic matter, making it available to other microorganisms. Bacteria, in turn, have developed defense mechanisms such as CRISPR-Cas systems to combat phage attacks (Maier et al., 2009; Murray et al., 2015). These dynamic interactions promote co-evolution, which improves microbial diversity and ecosystem stability.

5.2.3. Fungi-Virus Interactions

Fungi can also host viruses called mycoviruses. Mycoviruses frequently exist asymptomatically within their fungal hosts, but they can affect fungal physiology and pathogenicity. For example, the mycovirus *Cryphonectria hypovirus* 1 reduces the virulence of the chestnut blight fungus *Cryphonectria parasitica*, resulting in a biological control method for the disease. These interactions highlight mycoviruses' ability to modulate fungal behavior and ecological impacts (Lievens et al., 2015).

5.2.4. Protozoa-Bacteria Interactions

Protozoa, particularly amoebae and ciliates, are important bacterial predators in both soil and water. Protozoa influence nutrient cycling by preying on bacteria. This predation pressure can cause the evolution of bacterial traits that improve survival, such as biofilm formation and toxin production. Furthermore, protozoa-bacteria interactions can aid in horizontal gene transfer (HGT) among bacteria because protozoa can serve as reservoirs and vectors for plasmids and other genetic elements (Maier et al., 2009).

5.2.5. Algae-Bacteria Interactions

In aquatic environments, algae and bacteria engage in complex interactions that can be mutualistic, commensal, or antagonistic. Algae produce organic carbon through photosynthesis, which provides nutrients to heterotrophic bacteria. In exchange, bacteria can provide algae with essential vitamins and degrade algal-derived organic matter, thereby promoting algal growth (Amin et al., 2015). However, some bacteria produce algicidal compounds that can lyse algal cells, affecting both algal blooms and primary production.

5.2.6. Archaea-Bacteria Interactions

Archaea and bacteria frequently coexist in extreme environments, such as hot springs, salt flats, and anaerobic sediments, interacting in a variety of ways. Syntrophic relationships are critical in methanogenic environments, as bacteria degrade complex organic compounds into simpler molecules that methanogenic archaea convert into methane (Amin et al., 2015). These interactions are critical for the efficiency of anaerobic digestion processes and the biogeochemical cycling of carbon.

5.2.7. Cross-Kingdom Interactions

Microbial interactions go beyond individual domains and involve cross-kingdom relationships. For example, the tripartite symbiosis between leguminous plants, nitrogen-fixing bacteria (rhizobia), and mycorrhizal fungi is an example of a complex mutualistic interaction that promotes plant growth and soil fertility. Similarly, lichens, which consist of fungi and photosynthetic partners (algae or cyanobacteria), have a mutualistic relationship in which the fungus provides a protective structure and the photosynthetic partner supplies organic carbon (Hurst et al., 2007; Maier et al., 2009).

5.2.8. Pathogenic Interactions

Bacteria, fungi, protozoa, and viruses are all pathogenic microorganisms that interact with their hosts in ways that can cause disease. Pathogenic bacteria, such as Salmonella and Staphylococcus, exploit host cells and evade immune responses, frequently forming biofilms that help them survive. To invade host tissues, fungal pathogens such as *Candida albicans* can switch between yeast and hyphal forms. *Plasmodium* spp., for example, has complex life cycles within its hosts that allow it to avoid immune detection and cause diseases like malaria (Maier et al., 2009; Amin et al., 2015). Viral pathogens, such as influenza and HIV, use the host's cellular machinery to replicate, frequently compromising the immune system.

5.3. NUTRIENT CYCLING

Microbial life forms play a crucial role in nutrient cycling, a fundamental process that maintains the balance of essential elements within ecosystems. Bacteria, fungi, viruses, protozoa, algae, and archaea each contribute uniquely to the cycling of carbon, nitrogen, phosphorus, sulfur, and other nutrients, driving ecological and biogeochemical processes that sustain life on Earth.

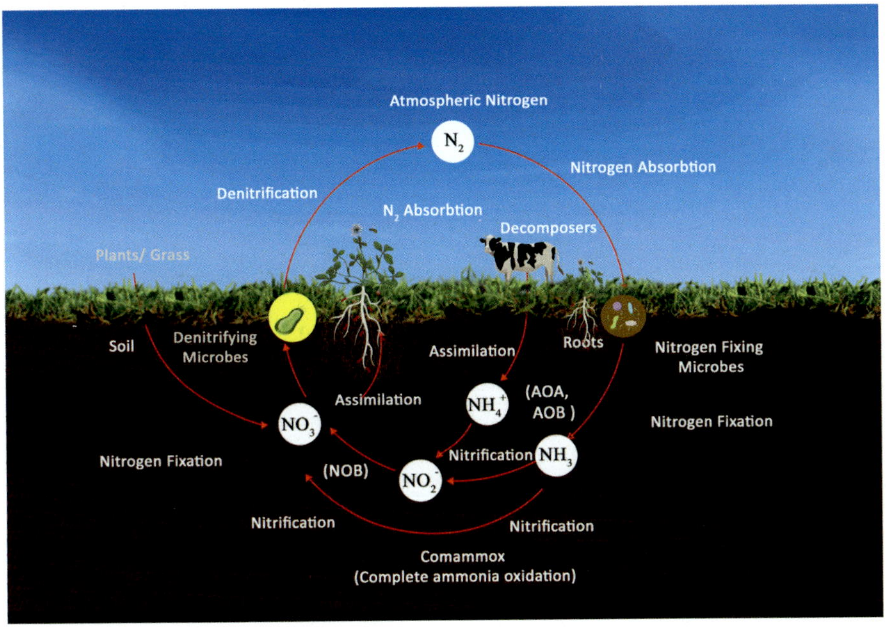

Figure 5.2. *The role of microorganisms in the nutrient cycling in agricultural soils.*

Source: Soratur A, Venmathi Maran BA, Kamarudin AS, Rodrigues KF. (2024). Microbial diversity and nitrogen cycling in peat and marine soils: A review. Microbiology Research, 15(2), 806-822. https://doi.org/10.3390/microbiolres15020052.

- **Bacteria:** Bacteria's metabolic diversity places them at the center of numerous nutrient cycles. In the carbon cycle, heterotrophic bacteria decompose organic matter, reducing complex organic compounds to simpler molecules like carbon dioxide and methane (Read and Perez-Moreno, 2003; Hurst et al., 2007). This decomposition is essential for carbon recycling in ecosystems. Autotrophic bacteria, such as cyanobacteria, contribute to primary production by fixing CO_2 through photosynthesis, converting inorganic carbon into organic forms that other organisms can use (Anderson, 2003).

Bacteria play important roles in the nitrogen cycle, including nitrogen fixation, nitrification, and denitrification. Nitrogen-fixing bacteria, such as Azotobacter and Rhizobium, convert atmospheric nitrogen (N_2) into ammonia (NH_3), which plants can assimilate. Nitrifying bacteria, such as Nitrosomonas and Nitrobacter, convert ammonia to nitrate (NO_3–), which plants can use more easily. Denitrifying bacteria, such as Pseudomonas and Clostridium, convert nitrate back into nitrogen gas (N_2), completing the nitrogen cycle and returning nitrogen to the atmosphere (Gougoulias et al., 2014).

- **Fungi:** Fungi are important decomposers in terrestrial ecosystems, breaking down complex organic matter, especially lignin and cellulose, which are resistant to bacterial degradation. Fungi like Aspergillus and Penicillium use enzymatic activities to release nutrients from dead plant and animal material, making them available to other organisms. Mycorrhizal fungi form symbiotic relationships with plant roots, increasing the surface area for absorption and mobilizing nutrients from the soil, particularly phosphorus (Maier et al., 2009; Vos et al., 2013).
- **Viruses:** Viruses, particularly bacteriophages, play an important role in microbial nutrient cycling by destroying bacterial cells. This process, known as the viral shunt, returns dissolved organic matter and nutrients to the environment for use by other microorganisms. The recycling of nutrients via viral lysis is critical for maintaining microbial population dynamics and nutrient availability in aquatic ecosystems (Anderson, 2003).
- **Protozoa:** Protozoa, as microbial predators, control bacterial populations by grazing. Protozoa consume bacteria and release nutrients in the form of waste products, which are then available for consumption by plants and other microorganisms (Read and Perez-Moreno, 2003). This grazing activity stimulates bacterial growth and activity, which improves nutrient mineralization and recycling throughout the ecosystem.
- **Algae:** Algae, as primary producers, are critical to the carbon and nitrogen cycles in aquatic environments. Algae use photosynthesis to convert carbon dioxide into organic carbon, forming the base of the food web and supporting higher trophic levels. Some algae, such as cyanobacteria, can fix nitrogen, which contributes to the nitrogen cycle by converting atmospheric nitrogen into forms that other organisms can use (Hurst et al., 2007; Maier et al., 2009).
- **Archaea:** Archaea, particularly those in extreme environments, play an important role in nutrient cycling. Methanogenic archaea in anaerobic environments, such as wetlands and ruminant guts, generate

methane from carbon dioxide and hydrogen, playing an important role in the carbon cycle. Nitrifying archaea, like *Nitrosopumilus maritimus*, convert ammonia to nitrite, a critical step in the nitrogen cycle. Furthermore, some archaea participate in sulfur cycling, converting sulfate to sulfide in anaerobic conditions (Vos et al., 2013).

5.4. MICROBIAL SUCCESSION

Microbial succession is the process by which a microbial community's composition and structure change over time within a specific environment. This process is influenced by a variety of biotic and abiotic factors and is critical for ecosystem development, nutrient cycling, and microbial community stability. Understanding microbial succession provides information about microbial ecology, biogeochemical cycles, and environmental adaptation (Maier et al., 2009).

5.4.1. Primary and Secondary Succession

There are two types of microbial succession: primary and secondary. Primary succession happens in previously uninhabited areas, such as newly exposed rock surfaces, volcanic deposits, or glacial retreats.

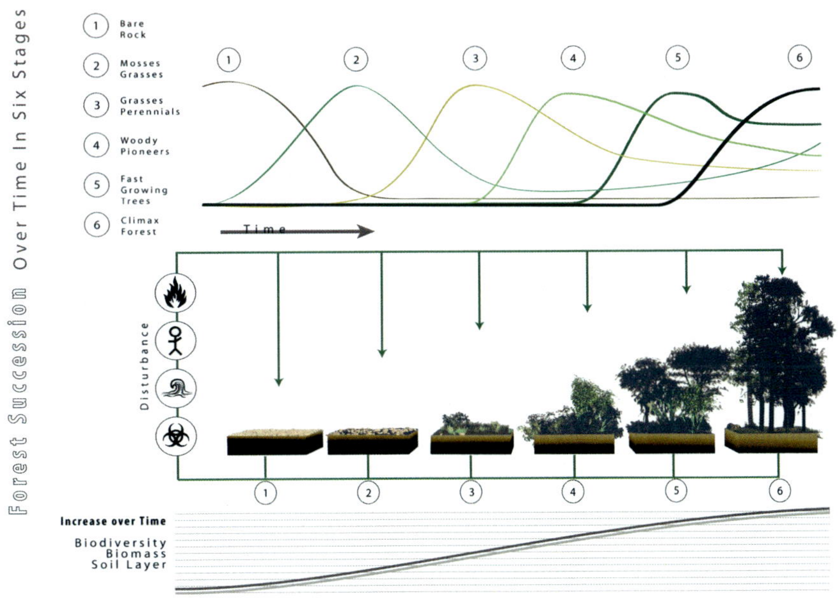

Figure 5.3. *Ecological succession describes stages of primary and secondary succession, leading to stable climax communities.*

Source: By LucasMartinFrey – Own work, CC BY 3.0, https://commons.wikimedia.org/w/index.php?curid=13274746.

In these environments, pioneer microorganisms like cyanobacteria and lichens are the first to colonize. These pioneers start soil formation by weathering rock surfaces and producing organic matter, which creates a favorable environment for future microbial communities (Fierer et al., 2010).

Secondary succession, on the other hand, occurs in disturbed environments that once supported microbial life. Examples include fallow agricultural fields, burned-out forests, and polluted ecosystems that are recovering. Secondary succession occurs when a microbial community re-establishes itself, often following a pattern influenced by the previous community's legacy and the nature of the disturbance (Ni et al., 2023).

5.4.2. Stages of Microbial Succession

The process of microbial succession typically follows several stages:

- **Pioneer Stage:** During primary succession, pioneer species such as autotrophic bacteria and cyanobacteria colonize the new environment. These microorganisms can fix carbon and nitrogen from the atmosphere, causing organic matter to form and contributing to soil development. Their metabolic activities alter the physical and chemical properties of the environment, making it more conducive to other microorganisms (Fierer et al., 2010).
- **Early Successional Stage:** After the pioneers, heterotrophic bacteria and fungi colonize the environment. These microorganisms decompose the pioneers' organic matter, releasing nutrients and altering the habitat. Early successional species frequently contain fast-growing, r-strategist organisms that can quickly exploit available resources (Hurst et al., 2007).
- **Mid-Successional Stage:** Microbial diversity grows in response to the complexity of the environment. Bacteria, fungi, protozoa, and other microorganisms make up the mid-successional community. Symbiotic relationships, such as mycorrhizal associations between fungi and plants, develop, improving nutrient availability and ecosystem stability (Maier et al., 2009; Vos et al., 2013).
- **Late Successional Stage:** In the late successional stage, the microbial community becomes more stable and mature. K-strategist organisms, which grow slowly and are better suited to stable environments, gain dominance. The microbial community structure grows more complex, with intricate interactions and nutrient cycling pathways emerging. Higher biodiversity and functional redundancy characterize the late successional stage, which contributes to ecosystem resilience (Fierer et al., 2010).

5.4.3. Factors Influencing Microbial Succession

Several factors influence microbial succession, including:

- **Environmental Conditions.** Temperature, pH, moisture, and nutrient availability have a significant impact on the composition and succession of microbial communities. Changes in these conditions can alter the competitive dynamics of microorganisms, causing shifts in community structure (Anderson, 2003; Gougoulias et al., 2014).
- **Disturbance Regimes.** Disturbances such as fire, flooding, or human activity can reset successional stages and change microbial communities. The frequency, intensity, and type of disturbance affect the trajectory of microbial succession (Ni et al., 2023).
- **Biotic Interactions.** Microorganism interactions, such as competition, predation, and symbiosis, are critical for succession. These interactions can drive the exclusion of specific species or promote the coexistence of diverse microbial groups (Maier et al., 2009; Fierer et al., 2010).

5.4.4. Ecological and Biogeochemical Implications

Microbial succession has important environmental and biogeochemical implications. It affects soil formation, nutrient cycling, and ecosystem productivity. For example, nitrogen-fixing bacteria's initial colonization in primary succession improves soil fertility, allowing plants and other organisms to establish themselves. Microbial succession in aquatic environments influences organic matter decomposition and element cycling, affecting water quality and ecosystem function (Vos et al., 2013; Gougoulias et al., 2014).

Understanding microbial succession also has practical implications for environmental management and restoration. It is possible to improve soil health, bioremediation, and agricultural productivity by manipulating successional stages and encouraging the establishment of beneficial microorganisms (Read and Perez-Moreno, 2003).

5.5. MICROBIAL COMMUNITIES

Microbial communities, which include a wide range of microorganisms, play critical roles in maintaining ecological balance, driving biogeochemical cycles, and supporting ecosystem function. Understanding these communities entails investigating their composition, interactions, and the environmental factors that influence them (Maier et al., 2009).

- **Bacterial Communities:** Bacteria are the most common and diverse microorganisms, appearing in almost every habitat on Earth. They form complex communities that vary greatly according to environmental conditions and available resources. Bacterial communities play an important role in soil ecosystems by cycling nutrients, decomposing organic matter, and promoting plant growth. For example, rhizosphere bacteria from the genera Pseudomonas and Bacillus improve nutrient availability while suppressing soil-borne pathogens (Maier et al., 2009).

Bacterial communities in aquatic environments help decompose organic matter and recycle nutrients. Marine bacterial communities, including those in the deep sea and coastal areas, participate in carbon and nitrogen cycling, which contributes to the overall productivity of marine ecosystems. The human gut microbiota, which includes a variety of bacterial species, such as Bacteroides and Firmicutes, is critical for digestion, immune function, and overall health (Callieri et al., 2019).

- **Fungal Communities:** Fungal communities play critical roles in decomposition and nutrient cycling, particularly in terrestrial ecosystems. Fungi, including decomposers like Penicillium and Aspergillus, degrade complex organic materials like cellulose and lignin, returning nutrients to the soil. Mycorrhizal fungi form symbiotic relationships with plant roots, enhancing water and nutrient uptake while protecting against pathogens (Vincent et al., 2021).

In forest ecosystems, ectomycorrhizal fungi from the genera Amanita and Boletus form extensive networks with tree roots, facilitating nutrient exchange and promoting tree growth. Endophytic fungi, which live within plant tissues without causing harm, help to maintain plant health by producing bioactive compounds that deter herbivores and pathogens (Hurst et al., 2007; Maier et al., 2009.

- **Viral Communities:** Viruses, particularly bacteriophages, are abundant and important members of microbial communities. They control bacterial populations through lysis, which promotes microbial diversity and evolution. In marine environments, viruses play an important role in controlling phytoplankton populations, which influences the marine food web and nutrient cycles. The viral shunt, a process in which viral lysis of microbial cells returns organic matter and nutrients to the environment, increases nutrient availability for other microorganisms (Vincent et al., 2021).

Viruses in the human microbiome, such as bacteriophages and eukaryotic viruses, interact with host cells and other microorganisms, influencing health and disease. The human virome, or collection of viruses within the human body, is a growing field of study with implications for disease understanding and therapeutic intervention development (Hurst et al., 2007; Callieri et al., 2019).

- **Protozoan Communities:** Protozoa are single-celled eukaryotes that play important roles in microbial communities, particularly as bacterial and microorganism predators. Protozoan grazing in soil and aquatic ecosystems regulates bacterial populations, which influences nutrient cycling and microbial community dynamics. For example, in freshwater environments, amoebae and ciliates help control bacterial abundance and contribute to organic matter breakdown (Maier et al., 2009).

Some protozoa are potent pathogens that cause disease in humans, animals, and plants. Plasmodium species, the malaria-causing agents, have complex life cycles that involve both vertebrate and invertebrate hosts, which has a significant impact on human health. Understanding protozoa communities and their interactions with other microorganisms is critical for disease management and ecosystem health (Vincent et al., 2021).

- **Algal Communities:** Algae are primary producers in aquatic ecosystems, forming the foundation of the food web and driving primary production through photosynthesis. Phytoplankton, which includes diatoms, dinoflagellates, and cyanobacteria, are essential components of marine and freshwater algal communities. They convert carbon dioxide into organic carbon, which benefits higher trophic levels and contributes to global carbon cycling (Callieri et al., 2019).

Algal blooms in marine environments, which are often caused by nutrient enrichment, can have serious ecological consequences. While some blooms promote productivity, others, such as those caused by harmful algal species, can emit toxins that harm marine life and human health (Anderson et al., 2012). Green algae and cyanobacteria are important in soil crusts and lichens in terrestrial environments, helping to fix nitrogen and stabilize the soil (Hurst et al., 2007).

- **Archaeal Communities:** Archaea are a distinct type of life that is commonly found in extreme environments such as hot springs, salt flats, and anaerobic sediments. They play critical roles in biogeochemical cycles, especially methane production and nitrogen cycling. Methanogenic archaea, such as those in the Methanococcus genus, produce methane from carbon dioxide and hydrogen in anaerobic environments, contributing to the global carbon cycle and greenhouse gas emissions (Callieri et al., 2019).

Ammonia-oxidizing archaea (AOA), such as *Nitrosopumilus maritimus*, contribute to the nitrogen cycle by converting ammonia to nitrite. Archaeal communities also interact with bacteria and other microorganisms, which helps microbial ecosystems stay stable and function properly (Hurst et al., 2007; Maier et al., 2009).

5.6. MICROBIAL BIODIVERSITY

Microbial biogeography, the study of the distribution of microorganisms across space and time, is a rapidly evolving field. Understanding how microbial communities are structured and function is pivotal in comprehending broader ecological and evolutionary processes. Below is a comprehensive discussion of some of the key areas in examining microbial diversity.

5.6.1. Distribution Patterns

A variety of factors influence microorganisms' dispersion patterns, resulting in various spatial distributions. Microbial communities have both universal and endemic distribution patterns. Certain bacteria and archaea are ubiquitous germs that can survive in a variety of environments (Martiny et al., 2006). In contrast, endemic microorganisms are limited to specific environments due to distinct local factors and historical contingencies.

High-throughput sequencing technologies have revealed intricate patterns of microbial diversity, demonstrating that while some microorganisms are international, others show strong biogeographical patterns impacted by both contemporary and historical processes (Martiny et al., 2006).

5.6.2. Environmental Factors

Environmental factors such as temperature, pH, salinity, and nutrient availability all influence microbial biogeography significantly. Soil pH, for example, plays an important role in determining the composition of bacterial communities in terrestrial environments. Similarly, salinity gradients in aquatic systems have a significant impact on microbial community structure, with separate communities adapted to freshwater, brackish, and marine environments (Lozupone and Knight, 2007).

Nutrient availability, such as carbon and nitrogen, also influences microbial dispersal since various microorganisms have distinct metabolic capacities and requirements. Furthermore, physical factors, such as soil texture and moisture content can produce microhabitats that support different microbial assemblages.

5.6.3. Ecological Interactions

Microbial biogeography is greatly influenced by ecological interactions that exist both between microorganisms and their surroundings, as well as among them. The shape and makeup of a community can be determined by competitive interactions for resources like space and nutrition. Diversity within a community can be shaped, for instance, by less competitive species being excluded due to microbial struggle for scarce nutrients (Martiny et al., 2006). By supplying certain habitats and resources, mutualistic interactions—such as those between nitrogen-fixing bacteria and plants—also significantly contribute to the structure of microbial communities. Furthermore, the regulation of microbial populations might affect their spatial distribution and community dynamics due to predation by bacteriophages and protozoa.

5.6.4. Evolutionary Processes

Microbial diversity and biogeography are determined by evolutionary mechanisms such as recombination, mutation, and HGT. Because of their enormous populations and short generation rates, microbes can swiftly adapt to changing environmental conditions, which leads to rapid evolution (De Mandal et al., 2022). For example, HGT makes it possible for new genetic features to be acquired quickly, promoting adaptation and survival in a variety of situations. Advantageous features, like antibiotic resistance, can spread throughout microbial communities as a result of this genetic exchange, which can take place between various microbial taxa. Modern microbial biogeography is also shaped by evolutionary history, which includes historical contingency and dispersal episodes. Previous events have a lasting impact on current distributions (Martiny et al., 2015).

5.6.5. Biogeography Theories

The distribution patterns of microorganisms have been explained by a number of biogeography concepts. According to the Baas-Becking theory, which states that "everything is everywhere, but the environment selects," environmental factors have the greatest influence on the spread of microorganisms (Martiny et al., 2006). According to this theory, although microbial dispersal occurs everywhere, some community compositions are chosen by local environmental conditions. On the other hand, the island biogeography theory, which was initially created for macro-organisms, has been modified to apply to microbial systems and highlights the significance of habitat isolation, size, and the dynamics of colonization and extinction in determining microbial diversity. A thorough framework for comprehending microbial biogeography is provided by metacommunity theory, which incorporates both local interactions and regional processes and takes into account both specific environmental conditions and more general spatial dynamics (De Mandal et al., 2022).

5.6.6. Applications

Numerous applications in environmental management, biotechnology, and human health stem from an understanding of microbial biogeography. Understanding the distribution of microorganisms might help environmental managers with bioremediation techniques, which use particular microbial populations to break down pollutants (Lozupone and Knight, 2007). Understanding the diversity and spread of microorganisms can help biotechnologists find new enzymes and bioactive substances with uses in industry and medicine. Researching the biogeography of dangerous bacteria can help us better understand how diseases spread and guide public health initiatives to prevent infectious diseases in humans. Additionally, because microorganisms are essential to the cycling of carbon and nitrogen, which affects ecosystem functioning and climate change, microbial biogeography advances our understanding of global biogeochemical cycles (Martiny et al., 2006).

5.7. MICROBIAL ECOSYSTEM SERVICES

The numerous services that microorganisms offer to human societies and ecosystems are referred to as microbial ecosystem services. These microorganisms, which include viruses, fungi, bacteria, and archaea, are essential to preserving the stability and well-being of both naturally occurring and artificially created ecosystems. They support vital functions such as soil formation, nutrient cycling, decomposition, and the control of plant and animal populations. It is essential to comprehend the functions of microbial ecosystems to promote environmentally friendly farming methods, environmental preservation, and public health. This section delves into the numerous vital services that microbes play in our ecosystem (Matulich et al., 2013).

5.7.1. Nutrient Cycling Services

The cycling of nutrients, which is essential to the sustainability and productivity of ecosystems, is greatly facilitated by microbes. They aid in the conversion and mobilization of vital nutrients like sulfur, phosphorus, and nitrogen. For example, nitrogen fixation is the process by which bacteria and archaea transform atmospheric nitrogen into ammonia, which plants can use (Cordell, 2000). According to Prosser et al. (2020), other bacteria take part in the processes of nitrification and denitrification, which control the availability of nitrogen and lessen its loss through leaching and gas emissions. Phosphate-solubilizing bacteria also mobilize phosphorus, increasing plant availability and boosting soil fertility (Rodríguez et al., 2018).

5.7.2. Bioremediation

Microbial bioremediation is an essential service that uses microorganisms' metabolic capacities to break down, convert, or detoxify environmental contaminants.

Various bacteria, fungi, and archaea are used to clean up contaminated habitats, such as soil and water bodies. For example, hydrocarbon-degrading bacteria are utilized to clean up oil spills by converting dangerous hydrocarbons into less toxic molecules (Das and Chandran, 2011). Heavy metal-resistant microorganisms can absorb and immobilize hazardous metals, lowering their bioavailability and environmental impact (Gadd, 2010). Bioremediation uses microbial variety and flexibility to repair contaminated habitats.

5.7.3. Soil Health and Fertility

Microbial communities exert a significant influence on the fertility and health of soil. Organic matter is broken down by soil bacteria, which release nutrients and strengthen the soil's structure.

By developing symbiotic associations with plant roots, mycorrhizal fungi improve soil aggregation and phosphate uptake. Furthermore, soil aggregation and stability—which are essential for halting erosion and preserving soil moisture—are facilitated by microbial exudates like polysaccharides (Lehmann et al., 2017). These microbial processes promote sustainable farming methods and increase soil fertility.

5.7.4. Plant-Microbes Interaction

Interactions between microbes and plants are crucial for the health and productivity of plants. Plant nutrient uptake and growth are facilitated by symbiotic connections, such as those that exist between nitrogen-fixing rhizobia and legumes (Graham and Vance, 2003).

By generating phytohormones, solubilizing nutrients, and fostering systemic resistance to infections, endophytic microbes—which live within plant tissues—can stimulate plant development (Matulich et al., 2013). Furthermore, by competing with pathogens for resources and space, generating antimicrobial chemicals, and triggering plant defenses, PGPR can decrease plant illnesses. The significance of microbial communities in plant resilience and agricultural productivity is highlighted by these interactions.

5.7.5. Biodiversity and Ecosystem Stability

The resilience and stability of ecosystems are largely dependent on microbial diversity. Diverse microbial populations contribute to functional redundancy to ensure that vital ecosystem functions can continue even in the face of stress or disruption. Rich microbial variety promotes the stability and productivity of ecosystems by supporting a range of ecological processes, including the breakdown of organic matter and the cycling of nutrients. Moreover, overall biodiversity and ecosystem health are promoted by microbial interactions with higher trophic levels, such as plants and animals. Therefore, maintaining microbial diversity is essential to keeping ecosystems strong and resilient (Matulich et al., 2013).

5.7.6. Climate Regulation

Microbes contribute significantly to climate regulation through their engagement in greenhouse gas cycles. Soil bacteria play an important role in the carbon cycle, facilitating the breakdown of organic matter and regulating carbon sequestration in soils. Methanogenic archaea create methane, a strong greenhouse gas, whereas methanotrophic bacteria oxidize it, reducing its release into the atmosphere. Furthermore, nitrifying and denitrifying bacteria control the generation and emissions of nitrous oxide, another strong greenhouse gas (Enwall et al., 2007). Through these mechanisms, bacteria help to regulate air composition and climate.

5.8. MICROBIAL BIOREMEDIATION

The term "bioremediation" describes the process of using living things, mostly microbes, to break down, purify, or eliminate toxins from the environment to return them to their natural state (Saeed et al., 2022). This method converts poisonous molecules into less toxic or non-toxic compounds by taking advantage of the metabolic powers of microbes. The process is an essential tool for controlling environmental contamination, especially in soils and water bodies, since it can happen naturally or be improved by human activity.

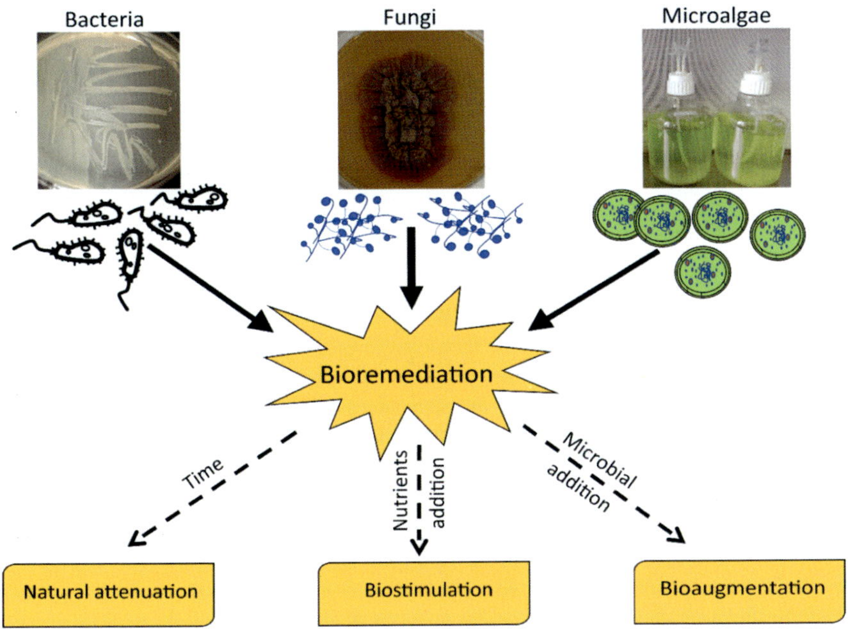

Figure 5.4. *Process of bioremediation.*

Source: Dell' Anno F, Rastelli E, Sansone C, Brunet C, Ianora A, Dell' Anno A. (2021). Bacteria, fungi and microalgae for the bioremediation of marine sediments contaminated by petroleum hydrocarbons in the omics era. Microorganisms, 9(8), 1695. https://doi.org/10.3390/microorganisms9081695.

5.8.1. Principles of Bioremediation

The principles underlying bioremediation are founded on the capacity of living organisms to utilize or transform pollutants.

- **Microbial Metabolism**: The metabolic activity of microorganisms, such as bacteria, fungi, and algae, is the fundamental component of bioremediation. Enzymes found in these species have the ability to convert complex organic contaminants into less hazardous, simpler forms. Anaerobic, facultative, or aerobic microbial metabolism can occur depending on the kind of pollutants and the surrounding conditions (Lederberg and Relman, 2009).
- **Pathways of Biodegradation**: Certain biodegradation routes, wherein microorganisms metabolize contaminants, are essential to the process of bioremediation. Under aerobic circumstances, organic pollutants are oxidized by microorganisms using oxygen, resulting in the production of carbon dioxide and water. Alternative electron

acceptors, like nitrates or sulfates, are employed in anaerobic settings to reduce pollution. The efficacy of bioremediation techniques is determined in large part by these mechanisms (Spang et al., 2015).

- **Bioaugmentation and Biostimulation**: These two techniques are essential to bioremediation practice. To improve the degradation of contaminants, foreign microbial strains with established degradative capacities are added in a process known as bioaugmentation. Contrarily, biostimulation entails altering the surrounding environment to stimulate the pre-existing native microbial populations and strengthen their inherent biodegradation mechanisms.
- **Optimization of the Environment**: Optimizing environmental factors like pH, temperature, and nutrient availability to promote microbial activity and development is necessary for effective bioremediation. Optimizing the settings for microorganisms involved in pollution degradation necessitates matching their unique requirements (Lederberg and Relman, 2009).

Thus, in order to treat environmental contamination, the fundamentals of bioremediation rely on utilizing the inherent metabolic capacities of microbes, improving environmental circumstances, and selectively implementing biotechnological interventions.

5.8.2. Factors Affecting Bioremediation

5.8.2.1. Environmental Factors

- **Temperature:** Temperature has a significant impact on microbial activity and, in turn, bioremediation rates. Pollutant breakdown is accelerated by optimal temperatures because they increase microbial metabolism and enzyme activity. Temperature extremes can hinder the activity of microorganisms or even kill them, which makes bioremediation less effective.
- **pH:** Enzymatic activity and microbial development are impacted by the environment's acidity or alkalinity. The majority of bioremediation techniques work best in a limited pH range. Variations from this range may result in toxicity to the implicated microorganisms or decreased microbial activity (Lederberg and Relman, 2009).
- **Oxygen Availability:** To support microbial metabolism and pollutant breakdown, aerobic bioremediation procedures need a sufficient supply of oxygen. Anaerobic reactions, on the other hand, require the presence of other electron acceptors. Limitations in oxygen

availability can seriously slow down the bioremediation process.

- **Nutrient Availability:** The availability of essential nutrients, such as nitrogen and phosphorus, can influence microbial growth and activity. Nutrient deficiencies can limit microbial proliferation and pollutant degradation efficiency (Saeed et al., 2022).

5.8.2.2. Biological Factors

- **Microbial Community Composition:** An essential component of bioremediation is the diversity and makeup of microbial communities. Certain bacteria have special enzymatic systems that can break down specific types of contaminants. Therefore, for bioremediation to be effective, competent microbial species must be present.
- **Genetic adaptability:** Microbes that have evolved to withstand environmental stressors or metabolic pathways unique to pollutants are better suited for bioremediation. Microbes can also be genetically engineered to increase or add certain properties, which will improve the breakdown of pollutants.

5.8.3: Types of Bioremediation

- **In Situ Bioremediation:** This technique manages the site's contaminated material. It involves methods such as air or oxygen injections into the soil to promote aerobic microbial activity (bioventing) and direct addition of particular microbial strains (bioaugmentation) to the site.
- **Ex Situ Bioremediation:** Material contaminated by contamination is taken out of the area to be treated. Techniques include biopiles, which include piling contaminated soil and periodically aerating it, and landfarming, which involves excavating contaminated soil and spreading it over a prepared bed for treatment (Saeed et al., 2022).

5.9. MICROBIAL CLIMATE CHANGE IMPACT

Microbial climate change impact investigates how changing climate conditions influence microbial communities and their functions. As temperatures rise and weather patterns change, microorganisms' distribution, activity, and ecosystem roles may fluctuate, affecting biogeochemical cycles and environmental health. Understanding these implications is critical for analyzing and managing climate change's effects on ecosystems.

5.9.1. Microbial Contribution to Greenhouse Gas Emissions

Microorganisms play a critical role in the Earth's biogeochemical cycles, particularly in the production and consumption of greenhouse gases including CO_2, CH_4, and N_2O. Microbial activities in soil, water, and sediments are major producers of these gases. Methanogenic archaea create methane by anaerobic digestion in wetlands, ruminant guts, and rice paddies, considerably adding to atmospheric CH_4 levels (Conrad, 2020). Nitrifying and denitrifying bacteria in the soil convert nitrogen molecules into nitrous oxide, a powerful greenhouse gas with a significantly larger global warming potential than CO_2 (Tiedje et al., 2022).

Microbes in marine habitats also help to regulate greenhouse gas levels. Photosynthesis by oceanic phytoplankton sequesters CO_2, while microbial respiration and breakdown release it back into the atmosphere. The balance of these processes impacts whether the ocean serves as a carbon sink or supplier. Thus, knowing microbial contributions to greenhouse gas emissions is critical for forecasting and mitigating climate change effects.

5.9.2. Effects of Climate Change on Microbial Communities

Microbial populations are greatly impacted by climate change because it changes the temperature, moisture content, and nutrient availability of their habitats. Warmer temperatures have the potential to quicken microbial metabolism, which raises the rate of breakdown and CO_2 emissions from organic materials in the soil (Tiedje et al., 2022). Additionally, warming modifies the nature of the microbial community, favoring thermophilic over mesophilic species, which can further affect greenhouse gas fluxes and ecosystem functioning (Singh et al., 2010).

Furthermore, variations in precipitation patterns influence soil moisture content, which in turn affects microbial activity and greenhouse gas emissions. For instance, wet conditions can promote anaerobic processes like denitrification and methanogenesis, boosting N_2O and CH_4 emissions, while drought conditions can decrease microbial activity and CO_2 emissions. Rising CO_2 levels cause ocean acidification, which affects marine microbial communities and may change the nutrient cycle and primary production (Tiedje et al., 2022). Thus, changes in microbial communities brought on by climate change have a significant impact on greenhouse gas dynamics and ecosystem health.

5.9.3. Microbial Interaction with Plants and Animals

Microbes have a substantial impact on ecosystem functioning and climate change effects. Symbiotic connections between microorganisms and plants, such as those between mycorrhizal fungi and nitrogen-fixing bacteria, promote plant growth and carbon sequestration. Mycorrhizal fungi, for example, improve plant nutrient intake and soil structure, allowing for greater carbon storage in soils (Tiedje et al., 2022).

In animal systems, gut bacteria are critical for digestion and nutrition cycling. Methanogens in ruminants' stomachs transform organic material into methane, which is then released into the atmosphere via enteric fermentation. Diet and environmental factors can impact the makeup and activity of gut microbiota, potentially impacting methane emissions from animals, a major source of anthropogenic CH_4 (Conrad, 2009).

Through pathogenic interactions, microbes also affect the health of plants and animals, and these interactions can be impacted by climate change. The spread and virulence of plant and animal infections can be increased by warmer temperatures and modified precipitation, which puts additional stress on ecosystems and reduces their capacity to absorb carbon (Conrad, 2009). In order to manage ecosystems and lessen the effects of climate change, it is essential to comprehend these relationships.

5.9.4. Microbial Feedback Mechanism

Microbial feedback mechanisms play an important role in climate change dynamics. Positive feedback loops can worsen climate change, while negative feedback loops can reduce it. Warming causes permafrost to thaw, releasing organic carbon that microbes degrade into CO_2 and CH_4. This increases the greenhouse effect and furthers warming. This positive feedback loop accelerates climate change and emphasizes the relevance of microbial activity in permafrost areas.

However, some microbial processes can generate negative feedback. Increasing CO_2 levels can boost plant growth and microbial activity, improving soil carbon sequestration and mitigating CO_2 emissions (Conrad, 2009). However, the overall impact of microbial feedback on climate change is complex and context-dependent, necessitating further research to understand and accurately forecast its consequences.

5.10. MICROBIAL CONSERVATION

Acknowledging the critical functions that microbial populations play in many ecosystems and human health, microbial conservation is an expanding field of study. Microbes, which include bacteria, fungi, viruses, and archaea, are incredibly small yet essential to biotechnology, biogeochemical cycles, and ecosystem processes. The dangers facing microbial communities, conservation tactics, and the significance of microorganisms for biotechnology and medicine are all covered in this talk.

5.10.1. Threats to Microbial Communities

Numerous threats impair the variety and functionality of microbial communities. Destroying habitats is one major threat. The natural ecosystems that support a variety of microbial communities are altered or destroyed by urbanization, deforestation, and agricultural growth. Microbial diversity is lost as a result of these activities, which also affect water quality, nitrogen cycling, and soil structure (Alonso, 2016).

Pollution represents yet another serious risk. Toxic chemical pollutants that can affect microbial communities include pesticides, heavy metals, and medications. These pollutants can suppress microbial development, change the makeup of communities, and lower metabolic activities that are vital to ecosystem services. Another serious threat to microbial diversity is climate change. The form and function of microbial communities can be changed by extreme weather events, changing precipitation patterns, and rising temperatures. For instance, higher temperatures have the potential to speed up microbial metabolism but also cause heat-sensitive species to disappear (Crowther et al., 2014).

Antibiotic-resistant bacteria have also emerged as a result of abuse of antibiotics, endangering human health in addition to upsetting microbial communities in their natural habitats. Resistance gene propagation among microbial populations has the potential to lower microbial diversity and impact the ecosystem services that these communities supply (Martinez, 2009).

5.10.2. Conservation Methods and Strategies

Diverse techniques and approaches are used in the conservation of microbial diversity. The preservation of natural environments is one strategy. The natural microbial communities that are found in environments including woods, wetlands, and oceans can be preserved with their protection. Protecting these ecosystems from human disruptions through conservation zones and protected reserves is essential (Martinez, 2009).

The use of sustainable farming methods is a further tactic. Soil health and microbial diversity can be improved by crop rotation, reducing the use of chemical fertilizers and pesticides, and implementing organic agricultural practices. By fostering a stable environment for helpful microorganisms, these methods support the cycling of nutrients and the health of plants.

In situ and Ex situ conservation techniques are also used. In situ, conservation preserves microbial communities in their native settings, whereas ex situ conservation keeps microbial cultures in controlled environments like culture collections and biobanks. These collections preserve microbial diversity while also providing resources for research and biotechnological applications (Maier et al., 2009).

Furthermore, the creation of microbial inoculants, which are preparations containing beneficial bacteria, has the potential to help rebuild microbial communities in degraded habitats. These inoculants can improve soil fertility, promote plant growth, and increase resistance to environmental stressors (Bashan et al., 2014).

5.10.3. Biotechnological and Medicinal Importance

The important functions that microorganisms play in biotechnology and medicine make microbial conservation imperative. Microbes are used in biotechnology because they can make important substances like enzymes, biofuels, and antibiotics. Microbes are employed in bioremediation to remove environmental contaminants, in fermentation to make food and drink, and in the production of bioenergy to create sustainable fuels (De Mandal et al., 2022).

Antibiotics, antivirals, and other medicines are derived from microorganisms for medical use. The discovery of penicillin, which was derived from the fungus *Penicillium notatum*, transformed medicine and demonstrated the use of germs in the creation of new medications. The search for novel molecules found in microbes that may treat a variety of illnesses, such as infectious disorders and cancer, is still ongoing (Maier et al., 2009).

Furthermore, a key factor in both health and sickness is the human microbiome, which is the collection of microorganisms that live in and on the human body. Digestion, immunity, and general health depend on a diversified and well-balanced microbiome. The relationship between disorders like obesity, diabetes, and inflammatory bowel disease and disruptions to the microbiome highlights the importance of microbial conservation concerning human health (De Mandal et al., 2022).

CHAPTER 6

MICROBIAL GENETICS

Contents

Microbial genetics is a subfield of microbiology that investigates the molecular mechanisms that govern heredity and variation in microorganisms, particularly bacteria. This field includes the study of genetic material, gene expression, genetic variation, and gene transfer between microbial populations. Understanding microbial genetics is critical for understanding how bacteria adapt to different environments, acquire antibiotic resistance, and evolve over time. Molecular tools and techniques, such as polymerase chain reaction (PCR) and next-generation sequencing (NGS), have significantly advanced the elucidation of bacterial genomes, revealing the complexity and dynamism of microbial genetic processes (Madigan et al., 2018).

The study of horizontal gene transfer (HGT) has revealed complex networks of genetic exchange that contribute to microbial diversity and innovation. Overall, microbial genetics provides important insights into the molecular basis of microbial life, with far-reaching implications for biotechnology, medicine, and environmental science (Moat et al., 2002).

6.1. BACTERIAL GENETICS

Bacterial genetics investigates the mechanisms of genetic variation, gene expression, and genetic material transfer in bacterial populations. This area of research is critical for understanding bacterial adaptation, evolution, and pathogenicity.

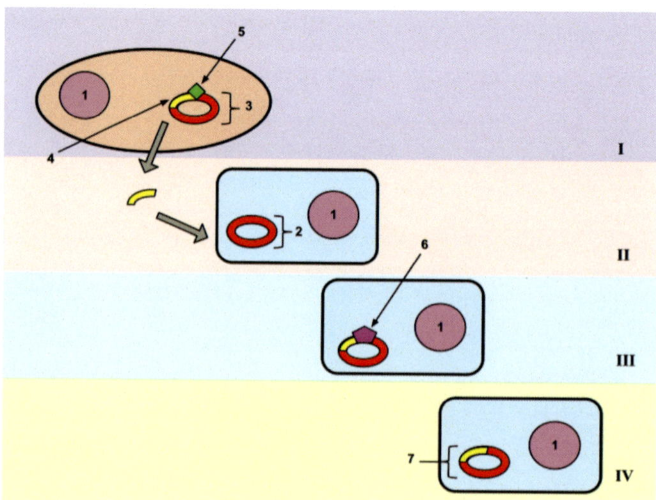

Figure 6.1. *Bacterial genetic transformation.*

Source: By Sprovenzano15 – Own work, CC BY-SA 3.0, https://commons.wikimedia.org/w/index.php?curid=26261580.

6.1.1. Genetic Material

Bacterial genetics studies the structure and function of genetic material within bacterial cells. The primary genetic material is typically a single, circular chromosome made up of double-stranded DNA. This chromosome contains essential genes that control the organism's vital functions, such as replication, metabolism, and structural integrity. Bacteria, unlike eukaryotic cells, do not have a nuclear membrane, so their DNA is located in the nucleoid. Aside from chromosomal DNA, bacteria frequently contain plasmids, which are smaller, circular DNA molecules that replicate independently of the chromosome. Plasmids frequently contain genes that confer beneficial characteristics, such as antibiotic resistance and virulence factors. The combination of chromosomal and plasmid DNA provides bacteria with the ability to adapt to a variety of environmental challenges (Murray et al., 2015).

6.1.2. Gene Expression and Regulation

Gene expression in bacteria entails the processes of transcription and translation, which convert genetic information into functional proteins. Transcription begins when RNA polymerase binds to a promoter region on the DNA, which starts the synthesis of messenger RNA (mRNA). Sigma factors regulate this process, directing RNA polymerase to specific promoters. Ribosomes are where transfer RNA (tRNA) molecules deliver amino acids following the mRNA codons (Baltz et al., 2010).

Bacterial gene regulation is frequently achieved through operons, which are clusters of genes transcribed as a single mRNA molecule. A classic example is the lac operon, which regulates lactose metabolism in *Escherichia coli*. The presence or absence of lactose regulates the lac operon by influencing the binding of repressor proteins to the operator region and thus modulating gene expression. Furthermore, regulatory proteins, such as repressors and activators, and small molecules, like cyclic AMP, play critical roles in fine-tuning gene expression in response to environmental signals (Hurst et al., 2007; Madigan et al., 2018).

6.1.3. Genetic Variation

Mutations and recombination generate genetic variation in bacterial populations. Mutations can occur naturally as a result of errors in DNA replication or as a result of external factors such as UV radiation and chemical mutagens. These mutations can be point mutations, insertions, deletions, or frameshift mutations, all of which have the potential to alter gene function and protein expression. Recombination, particularly homologous recombination, allows for the exchange of genetic material between similar DNA sequences, resulting in genetic diversity (Barton, 2010). Transposable elements, or transposons, contribute to genetic diversity by moving within and between genomes, potentially disrupting genes or regulatory regions (Baltz et al., 2010; Murray et al., 2015).

6.1.4. Horizontal Gene Transfer

HGT is an important mechanism for bacterial adaptation and evolution, allowing for the acquisition of new genetic traits from other organisms. HGT takes place through three main processes: transformation, transduction, and conjugation. Griffith's experiment with *Streptococcus pneumoniae* demonstrated that transformation involves the uptake of naked DNA from the environment. Bacteriophages facilitate transduction by transferring DNA between bacterial cells. Conjugation necessitates direct cell-to-cell contact and is frequently aided by plasmids, such as the F plasmid in *E. coli*, which contains genes for pilus formation and DNA transfer. These mechanisms allow bacteria to rapidly acquire and spread beneficial genes, including those that confer antibiotic resistance and virulence (Hurst et al., 2007).

6.1.5. Molecular Tools and Techniques

Advances in molecular tools have greatly improved the study and manipulation of bacterial genetics. PCR is a fundamental technique for amplifying specific DNA sequences, which allows for more detailed genetic analyses and diagnostics (Mullis et al., 1986). Gel electrophoresis separates DNA fragments based on their size, allowing for the visualization and characterization of genetic material. DNA sequencing technologies, such as NGS, offer in-depth insights into bacterial genomes, revealing genetic variations, pathogenic factors, and evolutionary relationships. CRISPR-Cas systems, initially discovered as bacterial immune mechanisms against phages, have been adapted for precise genome editing, allowing targeted modifications of genetic material (Moat et al., 2002).

These molecular tools hold significant implications for biotechnology and medicine. For example, genetically engineered bacteria are used to produce pharmaceuticals like insulin and antibiotics, as well as in environmental applications like bioremediation (Stephanopoulos, 2007). Furthermore, understanding the genetic basis of antibiotic resistance and developing novel antimicrobial strategies are critical to addressing the global threat of multidrug-resistant bacterial strains (Madigan et al., 2018, Murray et al., 2015).

6.2. FUNGAL GENETICS

Fungal genetics studies the hereditary material and genetic mechanisms of fungi, a diverse group of eukaryotic organisms. This field of study is critical to understanding the biology, ecology, and evolution of fungi, as well as their roles in human health, agriculture, and industry.

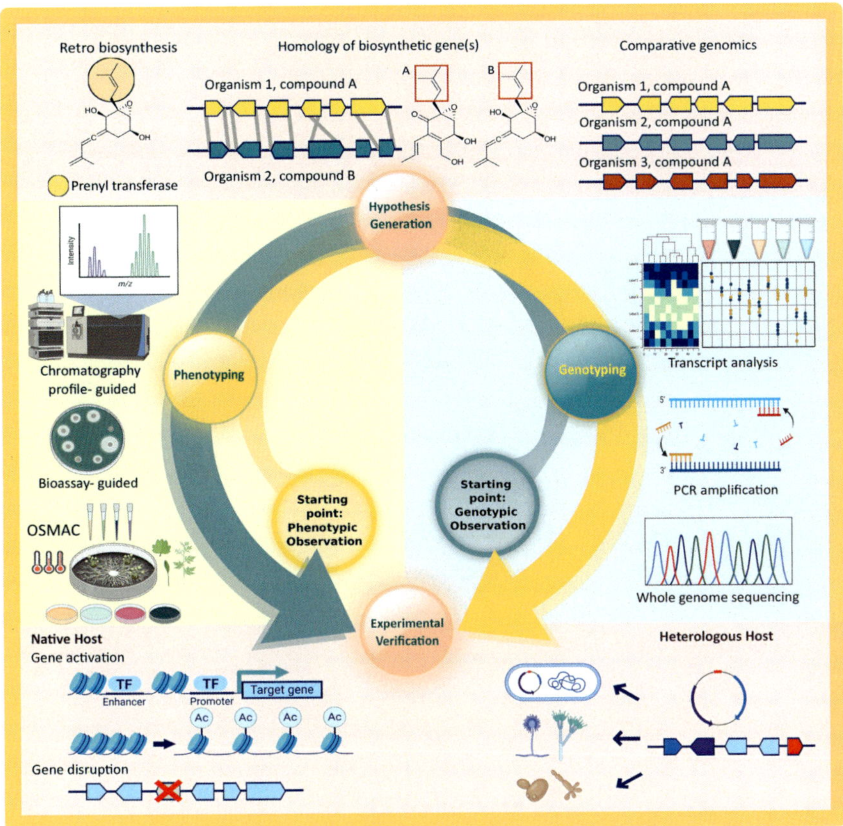

Figure 6.2. *Decoding genetic pathways in fungi.*

Source: Sagita R, Quax WJ and Haslinger K. (2021) Current state and future directions of genetics and genomics of endophytic fungi for bioprospecting efforts. Front. Bioeng. Biotechnol., 9, 649906. doi: 10.3389/fbioe.2021.649906.

6.2.1. Genetic Material

Fungi have multiple linear chromosomes contained within a nucleus. The size and complexity of fungal genomes vary greatly; for example, the yeast *Saccharomyces cerevisiae* has approximately 12 million base pairs spread across 16 chromosomes, whereas the filamentous fungus *Neurospora crassa* has approximately 40 million base pairs. In addition to nuclear DNA, fungi have mitochondrial DNA, which encodes genes necessary for energy production and metabolic processes. Fungal chromosomes are associated with histones and undergo chromatin remodeling, which affects gene expression and genome stability (Murray et al., 2015).

6.2.2. Gene Expression and Regulation

Fungi use the standard eukaryotic gene expression paradigm, which includes transcription and translation. Transcription occurs in the nucleus, where RNA polymerase converts DNA to messenger RNA (mRNA). Promoter regions, transcription factors, and enhancers control this process, which can either upregulate or downregulate the expression of specific genes. Splicing, capping, and polyadenylation are examples of post-transcriptional modifications required for mRNA stability and cytoplasmic export. Ribosomes in the cytoplasm decode mRNA to synthesize proteins (Moat et al., 2002; Baltz et al., 2010). Environmental cues, developmental stages, and cellular needs all influence fungal gene expression. One well-studied regulatory mechanism is the regulation of gene expression in response to nutrient supply. In S. cerevisiae, for example, the GAL4 transcription factor regulates the GAL genes involved in galactose metabolism, which activates transcription when galactose is present and represses transcription when it is not. In fungi, epigenetic modifications such as DNA methylation and histone modifications play important roles in regulating gene expression (Murray et al., 2015).

6.2.3. Genetic Variation

Mutations, sexual reproduction, and recombination all contribute to fungi's genetic diversity. Mutations can occur naturally as a result of errors in DNA replication or as a result of environmental factors such as UV radiation and chemical mutagens. These mutations can be point mutations, insertions, deletions, or frameshift mutations, all of which have the potential to affect gene function and phenotype (Hurst et al., 2007). In fungi, sexual reproduction begins with the fusion of haploid cells to form a diploid zygote, followed by meiosis, which produces genetic diversity through recombination and independent chromosome assortment. For example, in *N. crassa*, sexual reproduction results in the formation of ascospores, each with a distinct genetic composition. Furthermore, parasexual cycles in some fungi, such as *Aspergillus nidulans*, include mitotic recombination and aneuploidy, which contribute to genetic diversity without sexual reproduction (Madigan et al., 2018).

6.2.4. Horizontal Gene Transfer

HGT is less common in fungi than in bacteria, but it is important for fungal evolution and adaptation. HGT refers to the acquisition of genetic material from unrelated organisms, which can result in new metabolic capabilities and adaptive advantages. Various fungal species have shown evidence of HGT, including the transfer of genes related to pathogenicity, secondary metabolism, and antibiotic resistance. For example, the rice blast fungus *Magnaporthe oryzae* has acquired genes from other fungi, increasing its virulence. HGT mechanisms in fungi include transformation, viral transduction, and conjugation, though these processes are less well understood than those in bacteria (Moat et al., 2002).

6.2.5. Molecular Tools and Techniques

PCR is a widely used method for amplifying specific DNA sequences, allowing for detailed genetic analyses and diagnostics. Gel electrophoresis and Southern blotting are used to separate and identify DNA fragments, whereas DNA sequencing technologies such as NGS provide comprehensive insights into fungal genomes (Baltz et al., 2010).

Transformation techniques, such as electroporation and *Agrobacterium*-mediated transformation, which introduce foreign DNA into fungal cells, make genetic manipulation of fungi possible. CRISPR-Cas9 technology has been adapted for precise genome editing in fungi, allowing for targeted genetic modifications to investigate gene function and regulation (Shan et al., 2018). These molecular tools have far-reaching implications for biotechnology, including the creation of fungal strains for industrial enzyme production, biofuel generation, and pharmaceuticals (Murray et al., 2015).

6.3. VIRAL GENETICS

Viral genetics is a specialized field that investigates viruses' genetic composition, expression, and evolutionary dynamics. This area of research is critical for understanding viral behavior, pathogenicity, and the development of antiviral therapies.

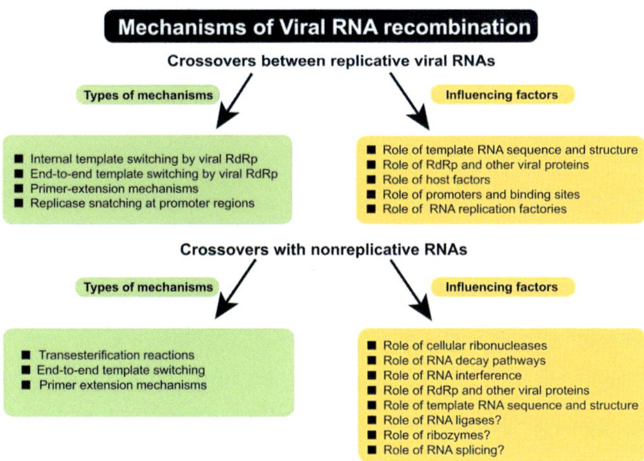

Figure 6.3. *Major mechanisms of virus genetic recombination.*

Source: Bujarski JJ (2013) Genetic recombination in plant-infecting messenger-sense RNA viruses: overview and research perspectives. Front. Plant Sci., 4, 68. doi: 10.3389/fpls.2013.00068.

6.3.1. Genetic Material

Viruses have genetic material that is made up of either DNA or RNA, which can be single-stranded (ss) or double-stranded (ds) and exist in linear or circular forms. Viral genomes are typically small, ranging from a few thousand to a few 100,000 base pairs, and contain essential proteins for viral replication and infection. The simplicity of viral genomes contrasts with the complex genomes of their host organisms. For example, the bacteriophage lambda has a dsDNA genome of about 48,500 base pairs, whereas the influenza virus has a segmented ssRNA genome of around 13,500 nucleotides (Madigan et al., 2018). These genomes are surrounded by a protein coat and, in some cases, a lipid envelope derived from the host cell membrane.

6.3.2. Gene Expression and Regulation

Viral gene expression is tightly controlled and linked to the host cell's machinery. DNA viruses usually use the host cell's transcription and translation systems to express their genes. Upon entry into the host cell, the host RNA polymerase transcribes viral DNA into mRNA, which is then translated into viral proteins. In contrast, RNA viruses frequently carry their own RNA-dependent RNA polymerase, which synthesizes mRNA from their RNA genomes. Retroviruses, such as HIV, reverse transcribe their ssRNA genomes into dsDNA, which is incorporated into the host genome and transcribed by the host machinery.

Regulation of viral gene expression involves various mechanisms, including promoter recognition, alternative splicing, and post-transcriptional modifications. For instance, the influenza virus utilizes a "cap-snatching" mechanism to hijack the host mRNA processing machinery (Moat et al., 2002).

6.3.3. Genetic Variation

Mutations, recombination, and reassortment are all mechanisms that cause genetic variation in viruses. The lack of proofreading mechanisms in RNA-dependent RNA polymerases causes RNA viruses to have high mutation rates. This results in the rapid evolution and adaptation of RNA viruses, as seen with the influenza virus and its frequent antigenic drift. Recombination occurs when two viral genomes infect the same cell and exchange genetic material, resulting in progeny with mixed genomes. This is common among DNA viruses and retroviruses. Reassortment, which is especially important for segmented RNA viruses like influenza, is the mixing of genome segments from different viral strains that co-infect a host, resulting in significant genetic shifts and the emergence of new viral strains (Hurst et al., 2007; Murray et al., 2015).

6.3.4. Horizontal Gene Transfer

Viruses are HGT agents, which transfer genetic material between species while also contributing to genetic diversity and evolution. Bacteriophages promote HGT in bacteria through transduction, which involves packaging bacterial DNA into phage particles and transferring it to new bacterial hosts. This can cause the spread of antibiotic resistance and virulence factors among bacterial populations. Retroviruses can insert their genomes into host DNA, occasionally transferring host genes between cells and organisms. Viral-mediated HGT is critical to the evolution of both viruses and hosts, influencing genetic innovation and adaptation (Baltz et al., 2010; Moat et al., 2002).

6.3.5. Molecular Tools and Techniques

Molecular tools and techniques have greatly aided the study and manipulation of viral genomes. PCR is a popular method for amplification and detection of viral genomes, allowing for precise genetic analysis and diagnostics. NGS technologies have transformed the field by revealing detailed information about viral genome sequences, genetic diversity, and evolution. Reverse genetics systems allow for the manipulation of viral genomes to investigate gene function and viral pathogenesis. For example, reverse genetics has been used to create recombinant influenza viruses, which will aid in the development of vaccines and antiviral strategies. CRISPR-Cas9 technology has also been used to target and edit viral genomes, potentially providing therapeutic applications for viral infections (Murray et al., 2015).

6.4. PROTOZOAN GENETICS

Protozoa are a diverse group of unicellular, eukaryotic organisms distinguished by their complexity and adaptability. Their genetic material, gene expression mechanisms, genetic variation, and genetic research tools all help to improve our understanding of their biology and pathogenesis.

6.4.1. Genetic Material

Protozoan genetic material is organized into linear chromosomes within a nucleus, indicating their eukaryotic nature. Protozoa's genome sizes vary greatly; for example, *Plasmodium falciparum*, the malaria parasite, has a genome of approximately 23 million base pairs distributed across 14 chromosomes, whereas *Trypanosoma brucei*, which causes African sleeping sickness, has a genome of about 26 million base pairs with 11 pairs of megabase chromosomes. Protozoan genomes also contain extranuclear DNA within organelles such as mitochondria, such as the kinetoplast DNA in trypanosomes, which is a distinct, highly structured network of circular DNA molecules (Madigan et al., 2018).

6.4.2. Gene Expression and Regulation

Protozoa express their genes by first converting DNA into mRNA in the nucleus and then translating it in the cytoplasm. However, protozoa have distinct regulatory mechanisms. *Plasmodium falciparum* gene expression is tightly linked to its complex life cycle, with distinct transcriptional patterns observed at various stages. Protozoan transcriptional regulation frequently involves promoter regions and transcription factors, but post-transcriptional mechanisms such as RNA interference (RNAi) and mRNA stability are also important. For example, in *Trypanosoma brucei*, double-stranded RNA induces the degradation of homologous mRNA, thereby controlling gene expression (Moat et al., 2002; Hurst et al., 2007).

6.4.3. Genetic Variation

Mutations, sexual recombination, and genetic exchange are some of the mechanisms that cause genetic variation in protozoans. Mutations occur as a result of errors in DNA replication or external mutagens, which contribute to genetic diversity. Although not all protozoa reproduce sexually, they do so through meiotic recombination and genetic exchange. For example, in *Plasmodium falciparum*, sexual reproduction occurs in the mosquito vector, resulting in genetic recombination. Furthermore, antigenic variation is prevalent in protozoa such as *Trypanosoma brucei*, where variant surface glycoprotein (VSG) genes switch frequently, allowing the parasite to evade the host immune system (Murray et al., 2015).

6.4.4. Horizontal Gene Transfer

HGT in protozoa is less well documented than in bacteria, but it is becoming more widely recognized as an important evolutionary mechanism. HGT enables protozoa to acquire genetic material from other organisms, which improves adaptability and survival. *Entamoeba histolytica*, for example, has acquired bacterial genes involved in metabolic pathways, which aids in its survival in anaerobic environments (Loftus et al., 2005). Similarly, evidence suggests that HGT has influenced the evolution of drug resistance in protozoan parasites (Baltz et al., 2010).

6.4.5. Molecular Tools and Techniques

Advances in molecular tools have significantly aided the study of protozoan genetics. The PCR amplifies specific DNA sequences, allowing genetic analysis and diagnostics. NGS technologies have transformed the field by revealing comprehensive information about protozoan genomes, transcriptomes, and epigenomes. CRISPR-Cas9 technology has been adapted for precise genome editing in protozoa, allowing for the manipulation of genes to study their function and develop new therapeutic strategies. Furthermore, RNAi-based techniques have been useful in functional genomics research in protozoa such as *Trypanosoma brucei* and *Plasmodium falciparum* (Murray et al., 2015).

6.5. ALGAL GENETICS

Algae, a diverse group of photosynthetic organisms, have a wide range of genetic characteristics that reflect their unique evolutionary histories and ecological niches. Understanding algal genetics is critical for realizing their full potential in biotechnology, ecology, and biofuel production.

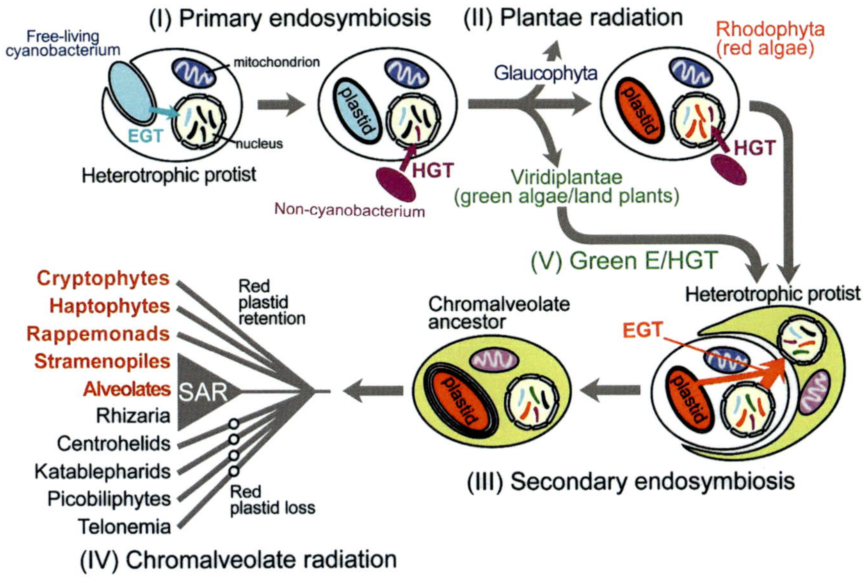

Figure 6.4. *Schematic illustration of cyanobacterial primary endosymbiosis and red algal secondary endosymbiosis that gave rise to the plastid in the vast majority of photosynthetic eukaryotes.*

Source: Qiu H, Yoon HS and Bhattacharya D (2013) Algal endosymbionts as vectors of horizontal gene transfer in photosynthetic eukaryotes. Front. Plant Sci., 4, 366. doi: 10.3389/fpls.2013.00366.

6.5.1. Genetic Material

The organization and content of genetic material in algae is extremely diverse. Green algae (*Chlorophyta*), red algae (*Rhodophyta*), and brown algae (*Phaeophyceae*) are all types of algae that can be unicellular or multicellular in nature. Their genetic material is located in the nucleus, chloroplasts, and mitochondria. Algae's nuclear genomes vary greatly in size and complexity; for example, the green alga *Chlamydomonas reinhardtii* has a relatively small genome of about 120 million base pairs, whereas the red alga *Porphyra umbilicalis* has a genome of about 105 million. Chloroplast genomes are typically circular, with genes required for photosynthesis and other chloroplast functions (Madigan et al., 2018).

6.5.2. Gene Expression and Regulation

Gene expression in algae is typical of eukaryotes, with DNA transcription into messenger RNA (mRNA) in the nucleus and protein translation in the cytoplasm. Several environmental factors influence gene expression in algae, including light, nutrient availability, and stress conditions. For example, in *Chlamydomonas reinhardtii*, light regulates the expression of genes involved in photosynthesis and circadian rhythm. Transcription factors, promoter regions, and enhancers all play important roles in controlling gene expression. Furthermore, post-transcriptional modifications and RNAi mechanisms are involved in the regulation of gene expression in algae (Moat et al., 2002).

6.5.3. Genetic Variation

Mutations, sexual reproduction, and recombination are all mechanisms by which algae acquire genetic diversity. Mutations can occur naturally or be caused by environmental factors like UV radiation and chemical mutagens. In algae, sexual reproduction occurs through the fusion of gametes, which results in genetic recombination and increased genetic diversity. For example, sexual reproduction in *Chlamydomonas reinhardtii* produces a zygote that goes through meiosis to produce genetically diverse offspring. Furthermore, HGT and endosymbiotic events have significantly contributed to algae genetic diversity, as evidenced by the presence of bacterial-derived genes in algal genomes (Hurst et al., 2007).

6.5.4. Horizontal Gene Transfer

HGT has contributed significantly to algae evolution and adaptation. This process has contributed to the development of new metabolic pathways and adaptation to changing environmental conditions. For example, the green alga *Chlamydomonas reinhardtii* has inherited genes from bacteria that improve its ability to metabolize various carbon sources. Similarly, the red alga *Porphyra umbilicalis* has incorporated bacterial genes to aid in nitrogen metabolism. These HGT events have had a significant impact on algae's metabolic capabilities and ecological success (Baltz et al., 2010; Madigan et al., 2018).

6.5.5. Molecular Tools and Techniques

PCR is a popular method for amplifying specific algal DNA sequences and performing detailed genetic analyses. Genetic transformation techniques, such as electroporation and *Agrobacterium*-mediated transformation, have been developed to introduce foreign DNA into algal cells, allowing for functional genomics research. NGS technologies have transformed algal genomics by revealing details about genome structure, gene content, and regulatory elements. Furthermore, CRISPR-Cas9 technology has been adapted for precise genome editing in algae, enabling targeted gene manipulation to study their functions and develop improved algal strains for biotechnological applications (Moat et al., 2002).

6.6. ARCHAEAL GENETICS

Archaeal genetics studies the genetic routes and traits unique to archaea, one of the three domains of life. Archaea, unlike bacteria and eukaryotes, have distinct genetic processes, including replication, transcription, and translation machinery. Understanding archaeal genetics expands our knowledge of fundamental biological processes, evolutionary history, and potential biotechnological applications.

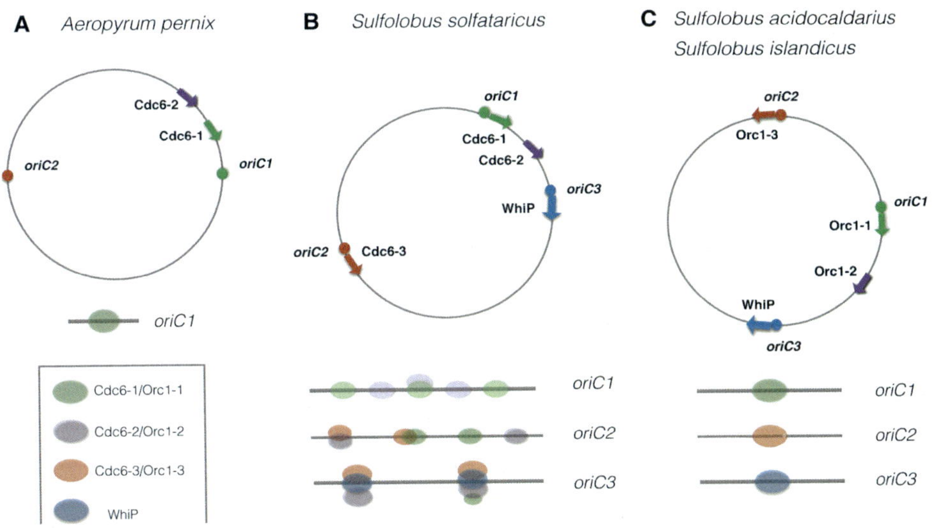

Figure 6.5. *Mechanisms of gene flow in archaea.*

Source: Ausiannikava D, Allers T. (2017). Diversity of DNA replication in the archaea. Genes, 8(2):56. https://doi.org/10.3390/genes8020056.

6.6.1. Genetic Materials

Similar to bacterial genomes, archaeal genomes are usually circular and comprise a single chromosome, while some archaea have several chromosomes or plasmids. High gene density and operon arrangement—groups of genes produced as a single mRNA strand—are common features of the genomic architecture of archaea. A distinct set of genetic characteristics that distinguish archaea from bacteria and remind them of both eukaryotic and bacterial systems. Among these are proteins that resemble histones and help to condense and arrange their DNA in a way that is similar to the chromatin structure of eukaryotes.

Additionally, archaeal genomes occasionally have introns, which are more typical of eukaryotic genes. According to Murray et al. (2015), the existence of these introns in archaea points to a complicated evolutionary past and the possibility of complex gene control mechanisms. The harsh conditions and diverse metabolic capacities of the archaeal domain are reflected in their huge genetic diversity (Murray et al., 2015).

6.6.2. Gene Expression and Regulation

A transcription mechanism that closely matches that of eukaryotes is involved in the expression of genes in archaea. TATA-binding proteins (TBPs) and transcription factor B (TFB) are used by archaeal RNA polymerases to start transcription, just like by the eukaryotic RNA polymerase II complex. These transcription factors detect the TATA boxes and BRE elements found in the promoter regions of archaea, much as they do in eukaryotes (Baltz et al., 2010; Moat et al., 2002).

Archaea are capable of regulating gene expression at three different levels: transcriptional, post-transcriptional, and translational. The Lrp family of proteins is an example of a transcriptional regulator that is important in controlling how genes are expressed in response to external stimuli. Additionally, heat shock proteins (HSPs) and other stress response components are upregulated in archaeal cells, which is one of their unique ways of coping with stress (Murray et al., 2015).

6.6.3. Horizontal Gene Transfer

HGT is an important step in archaeal evolution because it allows for the acquisition of genetic material from both bacterial and archaeal donors. This mechanism has considerably increased archaea's genetic variety and flexibility, allowing them to live in severe settings. Archaea use HGT mechanisms such as transformation, transduction, and conjugation. Studies have indicated that gene transfer from bacteria to archaea occurs more frequently than previously anticipated, indicating a significant interdomain genetic interchange. Many metabolic genes in archaea appear to be bacterial in origin, acquired through HGT processes. These genes frequently bestow favorable features, like as novel metabolic pathways or stress resistance mechanisms, which help the recipient archaea survive and thrive in their environment. Furthermore, the existence of integrative and conjugative elements (ICEs) in archaeal genomes strengthens the importance of HGT in their genetic history (Moat et al., 2002).

6.6.4. Molecular Tools and Techniques

The development of molecular tools and procedures has greatly accelerated the research of archaeal genetics. Genetic manipulation in archaea has always proven difficult due to their distinct cell wall architectures and resistance to standard antibiotics employed in bacterial systems. However, recent discoveries have brought a variety of genetic systems and methods designed specifically for archaea, such as transformation protocols, selectable markers, and gene knockdown techniques (Shan et al., 2018).

CRISPR-Cas systems, first identified in archaea, have been used as effective genome editing tools. These technologies allow for exact alterations to the archaeal genome, which facilitates functional genomics research and gene function discovery. Furthermore, advances in sequencing technology have enabled extensive genomic investigations, providing insights into archaeal species' evolutionary links and functional capacities (Baltz et al., 2010).

6.7. GENETIC VARIATIONS IN MICROORGANISMS

Microorganisms' adaptability, survival, and evolution are all heavily influenced by genetic variation. Mutations, gene transfers, and recombination processes all contribute to these variances, which result in a variety of phenotypes and capacities. Understanding these genetic changes is critical for investigating microbial resistance, pathogenicity, and ecological interactions. This information can be used to improve public health initiatives, treatment approaches, and biotechnological applications, emphasizing the importance of genetic study in microorganisms.

6.7.1. Types of Genetic Variations

Microorganisms depend on their genetic diversity to enable them to adapt and survive in a variety of settings. These changes result from variations in their DNA's nucleotide sequences. Mutations, genetic recombination, and HGT are the main forms of genetic diversity.

- **Mutations:** are alterations in the DNA sequence that can happen on their own or as a result of outside influences like UV rays or chemical mutagens. Point mutations, insertions, deletions, and frameshift mutations are among them; all have the ability to change how genes function.
- **Genetic Recombination**: New genetic combinations can be created through genetic recombination, which is the exchange of genetic material between distinct DNA molecules. This mechanism is very common in bacteria and viruses, where it can arise from site-specific recombination or homologous recombination.
- **Horizontal Gene Transfer**: The movement of genetic material between organisms other than by vertical transmission (from parent to offspring) is known as horizontal gene transfer or HGT. Three different methods can cause HGT: conjugation (transfer by direct cell-to-cell contact), transduction (transfer by bacteriophages), and transformation (uptake of naked DNA). In prokaryotes, HGT plays a major role in determining genetic diversity and adaptability (Baltz et al., 2010; Moat et al., 2002).

6.7.2. Mechanisms Driving Genetic Variations

The processes underlying genetic differences are intricate and multidimensional. They comprise interactions with the outside world as well as internal activities that bacteria go through.

- **DNA Replication**: Genetic mutations are mostly caused by mistakes in DNA replication. Even while DNA polymerases are capable of proofreading, mistakes can still happen and result in point mutations or more significant sequence alterations.
- **Recombination Mechanism**: Genetic material can be exchanged and rearranged through recombination mechanisms like homologous recombination. Although bacteria also use recombination during transformation and conjugation, this is especially important during processes like meiosis in eukaryotic microorganisms (Madigan et al., 2018).
- **Transposable Elements**: Jumping genes, or transposable elements, are able to travel about the genome and alter its genetic structure by adding, removing, or rearranging. Both genomic stability and gene expression may be significantly impacted by these factors.
- **Environmental Factors**: Genetic alterations can be brought about by environmental variables like radiation, toxins, and physical settings. For example, ultraviolet (UV) light can induce thymine

dimers in DNA, which, if left unchecked, might result in mutations. The creation of resistant strains of microbes can be attributed to selective forces in the environment, such as antibiotics, which favor genetic changes that are advantageous to the organism (Moat et al., 2002; Hurst et al., 2007).

6.7.3. Techniques for Studying Genetic Variations

Studying genetic variations in microorganisms requires a variety of molecular and computational techniques.

1. PCR is a fundamental technique that allows for the amplification and analysis of specific DNA sequences, facilitating the detection of mutations and genetic variations.
2. DNA Sequencing Technologies, notably NGS, have transformed the study of genetic variation. NGS enables the full study of entire genomes, finding mutations, insertions, deletions, and other changes with high resolution.
3. Microarrays allow for the simultaneous investigation of thousands of genomic sequences, yielding insights into gene expression patterns and genetic diversity within populations.
4. CRISPR-Cas9 technology enables precise editing of genetic material, making it an effective tool for investigating the functional implications of certain genetic variants.
5. Bioinformatics tools and software are critical for evaluating and interpreting the massive volumes of data produced by sequencing and other molecular procedures. These techniques assist detect genetic differences, forecasting their effects, and comprehending their roles in microbial evolution and adaptability (De Mandal et al., 2022).

6.7.4. Applications and Implications of Genetic Variations

The study of genetic differences in microorganisms has wide-ranging applications and ramifications in sectors including medicine, agriculture, and the environment.

- **Medical Use:** Understanding the processes of antibiotic resistance, a major public health concern is one of their medical uses. Identifying genetic variants associated with resistance allows for the development of new ways to combat resistant strains. In addition, genetic changes in infections can help with vaccine development and targeted therapeutics (Murray et al., 2015).

- Agricultural applications entail using genetically varied microbial populations to improve soil health, increase crop yields, and create biocontrol agents for pests and diseases. Understanding genetic variations in symbiotic bacteria can also lead to more efficient usage of biofertilizers.
- Bioremediation, which uses genetically varied microbes to break down toxins and recover damaged ecosystems, is one example of an environmental application. Genetic variation analysis can aid in identifying strains with improved degradative capacities.
- Genetic variation drives microorganism adaptability and speciation, which has profound evolutionary and ecological repercussions. This insight can help to guide conservation efforts and microbial community management in both natural and manmade habitats (Murray et al., 2015).

6.8. HORIZONTAL GENE TRANSFER

HGT is the flow of genetic material across organisms in a way other than vertical transmission (parent to offspring). Unlike traditional methods of genetic inheritance, which entail the transmission of genes within a lineage, HGT allows for the exchange of genes between species, promoting fast genetic variety and adaptability. This process is especially common in prokaryotes, such as bacteria and archaea, although it can also occur in eukaryotes. HGT is important in organismal evolution because it promotes genetic innovation and the spread of favorable features.

6.8.1. Types of Horizontal Gene Transfer

There are three main types of HGT:

1. **Transformation:** The uptake of naked DNA fragments from the environment by a competent bacterial cell. The foreign DNA can be absorbed into the recipient's genome, causing genetic alterations.
2. **Transduction:** This type happens when bacteriophages (viruses that infect bacteria) transfer genetic material from one bacterium to another. During infection, a bacteriophage may unintentionally wrap the host bacterial DNA and transmit it to another bacterial cell.
3. Conjugation is the direct transfer of DNA from one bacterial cell to another via a physical link known as a pilus. Plasmids are small, circular DNA molecules that can replicate independently of chromosomal DNA and are commonly used in conjugation (De Mandal et al., 2022).

6.8.2. Process and Mechanism of Horizontal Gene Transfer

- **Transformation:** Bacteria can take up exogenous DNA from their surroundings under specific conditions. The process involves DNA attaching to the cell surface, which is then taken up and integrated into the host genome. Bacterial competence, or the ability to take up DNA, can be natural or produced. Natural competence occurs in particular bacterial species and is controlled by certain genes, whereas artificial competence can be created in the lab using chemical or physical means.
- **Transduction:** Bacteriophages mediate transduction. It is possible for bacterial DNA to inadvertently be packaged into phage particles during a lytic cycle. The DNA that these phage particles carry can be reassembled into the genome of the new host when they infect another bacterium. Transduction comes in two flavors: specialized and generalist. Whereas specialized transduction only transfers particular genes close to the phage integration site, generalized transduction can transfer any bacterial gene.
- **Conjugation:** Conjugation is the transfer of plasmids via a mating bridge created by a pilus. The donor cell must carry a conjugative plasmid, such as the *F plasmid in E. coli*, which encodes the machinery required for pilus formation and DNA transfer. During conjugation, a single strand of plasmid DNA is transferred to the recipient cell and reproduced, resulting in a double-stranded plasmid (Boto, 2010).

6.8.3. Impact on Microbial Evolution

Microbial evolution is greatly impacted by HGT because it makes it possible for new features to be acquired quickly, which can be beneficial in environments that are changing. Through the transfer of genes linked to metabolic capacity, virulence factors, and antibiotic resistance, this genetic exchange helps microbial populations remain adaptable and survive. HGT has the potential to obfuscate species boundaries, resulting in mosaic genomes and making phylogenetic analysis more difficult. Additionally, it increases genetic diversity across microbial populations, which facilitates niche adaptability and ecological interactions (Boto, 2010).

6.8.4. Role in Pathogenicity and Anti-Microbial Resistance

HGT contributes significantly to the development of pathogenicity and antibiotic resistance in bacteria. Pathogenic bacteria can acquire virulence genes from other organisms, which increases their ability to cause disease. For example, transferring toxin genes via plasmids or bacteriophages can transform a non-pathogenic bacterium into a virulent pathogen. Similarly, the transmission of antibiotic-resistant genes via HGT poses a significant public health risk. Resistance genes can be transferred between bacterial species, resulting in the creation of multi-drug-resistant strains that are challenging to treat (Lederberg and Relman, 2009).

6.8.5. Detection and Study of Horizontal Gene Transfer

A combination of molecular biology, genomics, and bioinformatics techniques are used in the detection and investigation of HGT. Through the use of comparative genomics, scientists can find genes that are unusual for a particular creature, indicating potential horizontal transfer. The existence and source of transferred genes can be verified using methods including PCR, sequencing, and hybridization. Sophisticated bioinformatics instruments examine genome sequences to identify indications of HGT, like atypical nucleotide composition or phylogenetic incongruence. Furthermore, transduction studies, transformation protocols, and conjugation tests are experimental techniques that shed light on the mechanisms and frequency of HGT (Murray et al., 2015).

6.9. GENETIC ENGINEERING TECHNIQUES

Genetic engineering, also known as genetic modification, is a set of technologies used to alter the genetic makeup of organisms. This manipulation involves the direct modification of DNA to achieve desired traits. Genetic engineering encompasses various techniques such as recombinant DNA technology, gene cloning, and CRISPR-Cas9 (Hewitt, 2004). These methods enable the insertion, deletion, or modification of genes within an organism's genome, facilitating advancements in biotechnology, agriculture, medicine, and research.

6.9.1. Historical Development

James Watson and Francis Crick discovered the DNA double helix structure in 1953, laying the groundwork for genetic engineering. After this achievement, scientists like Paul Berg, Herbert Boyer, and Stanley Cohen developed recombinant DNA technology in the 1970s. This era saw the development of the ability to clip and paste DNA from various creatures, resulting in the creation of genetically modified organisms (GMOs).

In the 1980s, Kary Mullis developed the PCR, which revolutionized molecular biology by permitting the amplification of specific DNA sequences. Genetically modified crops, such as Bt corn and Roundup Ready soybeans, were initially introduced into agriculture in the 1990s. The 21st century has seen the advent of CRISPR-Cas9, a game-changing gene-editing technology that has simplified the process of targeted genetic modification and has wide-ranging applications in many industries (Murray et al., 2015).

6.9.2. Process and Techniques of Genetic Engineering

The process of genetic engineering consists of several critical steps:

- **Isolation of the Gene of Interest**: This entails identifying and isolating the exact gene that will be changed or introduced. Gene amplification is accomplished using techniques such as PCR.
- **Insertion of the Gene into a Vector**: The isolated gene is inserted into a vector (a carrier DNA molecule), such as plasmids or viruses, to ease gene transfer into the host organism.
- **Introduction to Host Organism**: The vector encoding the desired gene is introduced into the host organism's cells using techniques such as transformation, transfection, or microinjection.
- **Selection and Screening**: Host cells that have successfully integrated the novel gene are discovered and selected. This frequently entails employing marker genes that provide antibiotic resistance or fluorescence.
- **Expression and Cloning**: The host organism is cultivated to express the new gene, and cloning procedures are utilized to replicate the GMO (Ōmura et al., 2001).

6.9.3. Applications of Genetic Engineering Techniques

Genetic engineering has vast applications across various fields:

- **Medicine:** Insulin, growth hormones, and vaccinations are produced by genetic engineering. It is also essential to gene therapy, which fixes faulty genes in an attempt to alleviate genetic illnesses.
- **Agriculture**: Increased crop yields and food security have resulted from the development of genetically modified crops with better features like pest resistance, herbicide tolerance, and greater nutritional value.
- **Industrial Biotechnology**: The creation of biofuels, bioplastics, and other bio-based materials is accomplished through the employment of genetically modified microbes.
- **Research**: Model organisms with certain genetic alterations may be created more easily through genetic engineering, which helps researchers better understand how genes function and how diseases develop (Boto, 2010).

6.9.4. Controversies and Ethical Issues

While genetic engineering has the potential to revolutionize medicine, agriculture, and biotechnology, it is riddled with controversy and ethical concerns. Concerns include the possibility of unexpected ecological consequences, such as disrupting natural ecosystems through the release of GMOs. Ethical debates revolve around the morality of modifying genetic material, particularly in humans, generating concerns about consent, equity, and the possibility of eugenics. Furthermore, there are concerns about biotech monopolies and the social consequences of private genetic technologies. The balance between innovation and ethical responsibility is still a heated and continuing debate in the field of genetic engineering (Moat et al., 2002).

6.10. APPLICATION OF MICROBIAL GENETICS

The study and modification of microbes' genetic material to comprehend their interactions, functions, and evolution is the application of microbial genetics. This field makes significant contributions to scientific and biotechnological advancements by exploring microbial diversity, addressing environmental challenges, developing novel therapeutics, and improving industrial processes through the use of advanced techniques like genome editing, sequencing, and gene cloning.

6.10.1. Medical Microbiology

The use of microbial genetics in medical microbiology has completely changed how we think about infectious diseases and how to design treatments. Microorganisms that have undergone genetic modification have made it possible to identify and characterize resistance genes, virulence factors, and pathogenic pathways. The precise editing and amplification of microbial genes has been made possible by methods like genome sequencing, PCR, and CRISPR-Cas9. This has allowed for in-depth research on the activities of these genes (Davies and Davies, 2010). For instance, the discovery of genes that confer antibiotic resistance in pathogenic bacteria has influenced the creation of novel antimicrobial approaches and diagnostic instruments, improving the treatment of infections that are resistant to drugs. Additionally, the production of recombinant vaccines, such as the hepatitis B vaccine, relies on the genetic engineering of microorganisms to express viral antigens, providing safe and effective immunization options (Murray et al., 2015).

6.10.2. Biotechnology and Industrial Technology

Microbial genetics is critical in biotechnology and industrial technology, enabling the creation of medications, biofuels, and enzymes. The use of genetically modified microorganisms (GMMs) to produce insulin, growth hormones, and other therapeutic proteins has considerably improved medical therapies (Hewitt, 2004). Microbial fermentation processes in industrial biotechnology are genetically engineered to increase yield and efficiency. For example, the genetic alteration of yeast and bacteria has increased bioethanol and biobutanol production, providing long-term alternatives to fossil. Furthermore, the creation of microbial cell factories that are engineered to create high-value substances such as antibiotics, vitamins, and biodegradable polymers emphasizes the importance of microbial genetics in industrial applications (Moat et al., 2002).

6.10.3. Agricultural Biotechnology

Microbial genetics is used in agricultural biotechnology to boost crop yield, shield plants against illness, and advance environmentally friendly farming methods. The application of GMMs as biopesticides and biofertilizers has demonstrated potential for enhancing plant health and soil fertility. For instance, the goal of introducing nitrogen-fixing genes into non-leguminous plants is to promote environmentally friendly agriculture by reducing reliance on artificial fertilizers (Beatty and Good, 2011). Furthermore, plant infections are managed by genetically engineered bacteria and fungi, which lessens the requirement for conventional pesticides. The importance of microbial genetics in agriculture is further demonstrated by the introduction of genetically modified plants, such as Bt cotton, that have increased resistance to pests and diseases (Enwall et al., 2007).

6.10.4. Food and Beverage Industries

The food and beverage industries rely heavily on microbial genetics to improve the quality, safety, and production of their goods. Probiotics with enhanced health benefits and functional features have been developed as a result of genetically modifying LAB, which are employed in the fermentation of dairy products (Sieuwerts et al., 2008). Genetic engineering of yeast strains has optimized fermentation processes in the brewing and wine-making sectors, enhancing alcohol content, flavor, and aroma. Microbial genomics has also made it easier to identify and manage foodborne infections, protecting public health and food safety. Methods like whole-genome sequencing and real-time PCR are used to track out the sources of contamination and stop epidemics (Olempska-Beer et al., 2006).

6.10.5. Environmental and Ecological Applications

Microbial genetics is applied in ecological and environmental contexts to tackle problems including waste management, pollution, and ecosystem restoration.

Bioremediation solutions for contaminated settings are provided by GEMs, which are designed to break down harmful pollutants such as plastics, heavy metals, and hydrocarbons. It is possible to improve soil fertility and lessen the effects of climate change by genetically modifying microorganisms involved in the carbon and nitrogen cycles. To increase the amount of nitrogen available in agricultural soils, for example, genes producing nitrogenase enzymes are inserted into non-nitrogen-fixing bacteria. Furthermore, by making it possible to monitor and manage microbial communities in diverse habitats, microbial genetics supports ecological balance and sustainability and aids in the conservation of biodiversity (De Mandal et al., 2022).

CHAPTER 7

MICROBIAL BIOTECHNOLOGY

Contents

Microbial biotechnology is a rapidly developing field that uses microorganisms' capabilities for a wide range of applications. This interdisciplinary science uses microorganisms' genetic, metabolic, and enzymatic properties to address issues in medicine, agriculture, industry, and environmental management.

One significant application is bioremediation, which uses microbes to detoxify pollutants and contribute to environmental sustainability (Malik et al., 2022). Microbial biotechnology is important in agriculture because it helps develop biofertilizers and biopesticides, which improve crop yield and disease resistance. Industrially, microorganisms are engineered to produce biofuels, bioplastics, and other valuable biochemicals, providing long-term alternatives to petrochemical processes (Baltz et al., 2010; Singh, 2019).

Advances in microbial genomics and synthetic biology have increased the potential of microbial biotechnology, allowing the generation of novel microbial strains with enhanced capabilities. Furthermore, the field contributes to healthcare through the development of antibiotics, vaccines, and probiotics, highlighting its importance to public health. As research advances, microbial biotechnology promises innovative solutions to global challenges, highlighting its importance in achieving sustainable development goals (Malik et al., 2022).

7.1. INDUSTRIAL MICROBIOLOGY

Microbiology is important in many industries because it uses the metabolic and genetic diversity of microorganisms to develop processes and products that are necessary in today's society. The field has made significant progress in recent years, thanks to technological advancements in areas such as genetic engineering, fermentation technology, and bioprocessing (Baltz et al., 2010).

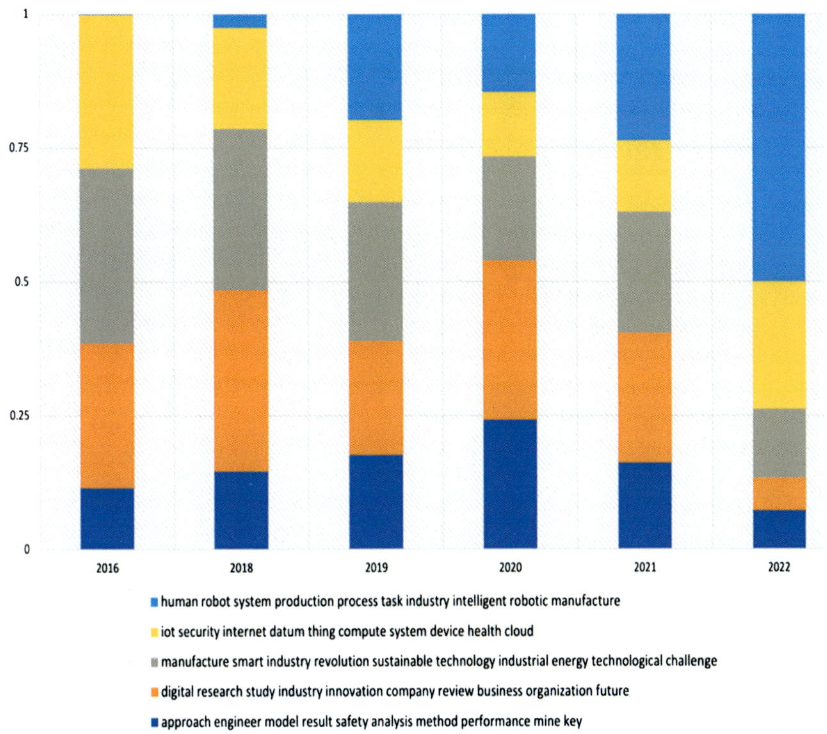

Figure 7.1. *Industrial microbiology market to grow significantly 2022–2027.*

Source: Akundi A, Euresti D, Luna S, Ankobiah W, Lopes A, Edinbarough I. (2022). State of industry 5.0—analysis and identification of current research trends. Applied System Innovation, 5(1), 27. https://doi.org/10.3390/asi5010027.

7.1.1. Biotechnology and Genetic Engineering

The development of genetic engineering has transformed the industrial applications of microbiology. Microorganisms, particularly bacteria and yeast, are engineered to produce a diverse range of products such as pharmaceuticals, biofuels, and enzymes. For example, *Escherichia coli* has been genetically modified to produce insulin, a vital drug for diabetes management. This application is an excellent example of how genetic engineering has enabled large-scale pharmaceutical production, lowering costs and increasing accessibility (Malik et al., 2022).

CRISPR-Cas9 technology, a recent breakthrough in genetic engineering, has improved the ability to edit microbial genomes precisely. This has resulted in the development of microorganisms with improved characteristics, such as higher yield or resistance to industrial stressors, increasing the efficiency of microbial processes in industries (Baltz et al., 2010).

7.1.2. Fermentation Technology

Fermentation is one of the oldest and most widely used industrial applications of microbiology. Microorganisms convert organic substrates into valuable products like alcohol, organic acids, and antibiotics. Beer, wine, yogurt, and cheese are all heavily reliant on fermentation in the food and beverage industry (Baltz et al., 2010).

Recent advances in fermentation technology include optimizing microbial strains and fermentation conditions to increase product yield and quality. For example, the development of high-performance yeast strains has greatly increased the efficiency of bioethanol production, a renewable energy source. Furthermore, advances in bioreactor design and process control have enabled the scaling up of fermentation processes, making them more viable for industrial applications (Malik et al., 2022).

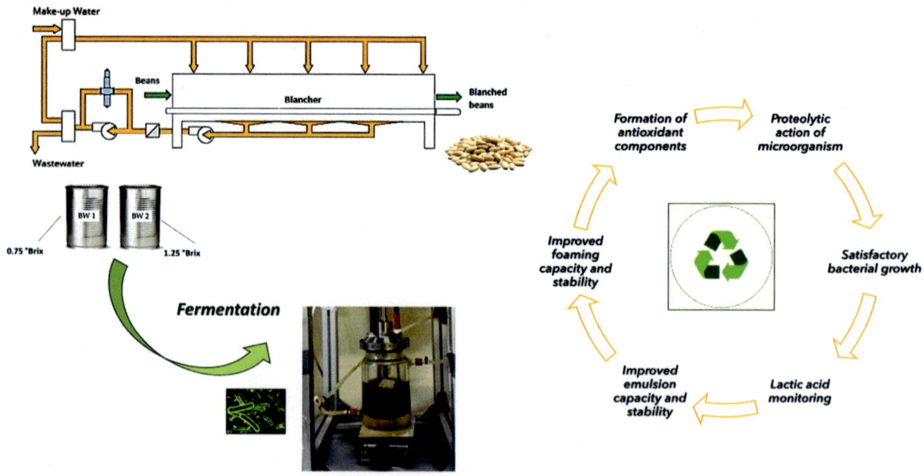

Figure 7.2. *Industrial fermentation process.*

Source: Colucci Cante R, Recupero A, Prata T, Nigro F, Passannanti F, Gallo M, Lentini G, Nigro R, Budelli AL. (2023). Valorization through lactic fermentation of industrial wastewaters from a bean blanching treatment. Fermentation, 9(4), 350. https://doi.org/10.3390/fermentation9040350.

7.1.3. Industrial Enzymes

Microorganisms are the primary source of industrial enzymes, which are used in a wide range of applications such as food processing, detergent production, and textile manufacturing. Bacteria and fungi produce enzymes such as amylases, proteases, and lipases, which are required for starch hydrolysis, protein degradation, and fat removal (Malik et al., 2022).

Recent technological advancements have centered on increasing the stability and activity of industrial enzymes in harsh environments such as high temperatures and extreme pH levels. Protein engineering and directed evolution are two techniques used to create enzymes with improved properties, increasing their industrial applications (Singh, 2019).

7.1.4. Biofuels and Bioplastics

Another important industrial application for microbiology is the production of biofuels and bioplastics. Microorganisms such as algae, bacteria, and yeast are used to convert renewable resources into biofuels such as biodiesel and bioethanol, which can be used instead of fossil fuels. Similarly, microbial fermentation processes using feedstocks such as sugar and plant oils produce bioplastics (Baltz et al., 2010).

Metabolic engineering advancements have enabled the optimization of microbial pathways for biofuel and bioplastic production, resulting in higher yields and lower costs. Furthermore, the advancement of synthetic biology tools has made it easier to design novel microbial strains capable of producing next-generation biofuels and bioplastics with superior properties (Malik et al., 2022).

7.2. AGRICULTURAL MICROBIOLOGY

Agricultural microbiology is the study and application of microorganisms in agricultural systems, with significant implications for crop production, soil health, pest control, and environmental sustainability. This interdisciplinary field uses microbial processes and technologies to increase agricultural productivity and resilience. Significant advances in this field have transformed modern agriculture, emphasizing the importance of microbiology in promoting sustainable agricultural practices (Rao, 2016).

7.2.1. Soil Microbiology and Fertility

Microorganisms contribute significantly to soil fertility, primarily through their roles in nutrient cycling. Soil bacteria, fungi, and actinomycetes decompose organic matter, releasing essential nutrients like nitrogen, phosphorus, and potassium that are required for plant growth. Nitrogen-fixing bacteria, such as Rhizobium spp., form symbiotic relationships with leguminous plants to convert atmospheric nitrogen into ammonia, a bioavailable form of nitrogen. This natural process reduces the need for synthetic nitrogen fertilizers and promotes sustainable agricultural practices (Singh, 2019).

7.2.2. Plant-Microbe Interactions

Plant-microbe symbiosis improves plant health and productivity. Mycorrhizal fungi, for example, form associations with plant roots, extending the root system's reach while improving water and nutrient uptake. These symbiotic interactions also help plants resist environmental stressors like drought and soil salinity. Pathogenic microorganisms, on the other hand, cause serious problems for plants. Advances in understanding plant immune responses and the development of microbial biocontrol agents have resulted in effective plant pathogen management strategies (Sammauria et al., 2020).

7.2.3. Biocontrol Agents

Biocontrol agents use beneficial microorganisms to control agricultural pests and diseases, providing an environmentally friendly alternative to chemical pesticides. *Bacillus thuringiensis*, for example, produces toxins that are lethal to certain insect pests, reducing the need for chemical insecticides. *Trichoderma* spp. are effective against a variety of plant pathogens, increasing disease resistance via mechanisms such as competition, antibiosis, and the induction of plant defense responses. The use of biocontrol agents is consistent with integrated pest management (IPM) strategies, which promote sustainable agricultural practices (Rao, 2016; Singh, 2019).

7.2.4. Microbial Inoculants

Biofertilizers and microbial inoculants add beneficial microorganisms to agricultural systems, improving soil fertility and plant health. *Azospirillum* spp. and *Azotobacter* spp., for example, are commonly used as biofertilizers to stimulate plant growth by fixing atmospheric nitrogen and synthesizing growth-promoting substances. The commercialization of microbial inoculants has gained traction, owing to the need for environmentally friendly agricultural inputs that reduce reliance on chemical fertilizers.

7.2.5. Genetic Engineering and Biotechnology

Advances in genetic engineering and biotechnology have increased the potential applications of microorganisms in agriculture. GMMs with enhanced traits, such as improved nitrogen fixation or pest resistance, present promising solutions to agricultural problems. The introduction of CRISPR-Cas9 technology has further transformed microbial genetics, allowing for precise genome editing to improve microbial functionalities tailored to specific agricultural applications (Singh, 2019; Malik et al., 2022).

7.2.6. Microbial Ecology and Sustainable Agriculture

Understanding microbial diversity and ecosystem dynamics is critical to developing sustainable agricultural practices. Crop rotation, reduced tillage, and organic amendments are examples of agroecosystem management practices that influence soil microbial communities and their functions (Sammauria et al., 2020). Promoting microbial diversity through sustainable practices improves soil resilience, nutrient cycling, and plant health, thereby contributing to long-term agricultural sustainability.

7.2.7. Food Safety and Post-Harvest Microbiology

Microbiological applications include food safety and post-harvest management, which ensures the quality and safety of agricultural products. Microbial contamination during cultivation, harvesting, and storage poses a serious threat to food safety. Advances in microbial detection and control methods, such as biopreservation with lactic acid bacteria (LAB), have increased food safety and agricultural product shelf life (Malik et al., 2022).

7.3. ENVIRONMENTAL MICROBIOLOGY

Microbiology is critical in addressing environmental issues, as it uses microorganisms to reduce pollution, manage waste, and restore ecosystems. Significant advances in environmental microbiology have resulted in the development of novel technologies and strategies, emphasizing the role of microbial processes in environmental sustainability.

- **Bioremediation:** Bioremediation is an important environmental application of microbiology that uses microorganisms to degrade or detoxify pollutants in soil, water, and air. Bacteria, fungi, and archaea have metabolic pathways that allow them to convert harmful contaminants into less toxic or nontoxic compounds. This process is especially effective for organic pollutants like hydrocarbons, pesticides, and solvents (Malik and Goyal, 2022).

One notable advancement in bioremediation is the use of GEMs that improve pollutant degradation capabilities. For example, *Pseudomonas putida* has been genetically modified to degrade aromatic hydrocarbons more efficiently, making it an effective solution for cleaning up oil spills and industrial waste (Maier et al., 2009). Furthermore, bioremediation strategies have been improved through bioaugmentation, which involves introducing specific microbial strains to polluted sites to speed up pollutant degradation.

- **Waste Management:** Microbial processes are essential for sustainable waste management, especially in the treatment of municipal and industrial waste. Microorganisms are used in wastewater treatment plants to degrade organic matter, lowering biochemical oxygen demand (BOD) and removing nutrients like nitrogen and phosphorus (Maier et al., 2009). To effectively treat wastewater, activated sludge systems rely on a diverse community of bacteria, protozoa, and metazoans.

Anaerobic digestion is another important application in which microorganisms decompose organic waste in the absence of oxygen, resulting in biogas (methane and carbon dioxide) and digestate (a nutrient-rich residue). This process not only reduces waste volume but also produces renewable energy, which contributes to a circular economy. Advances in metagenomics and bioinformatics have improved our understanding of the microbial communities involved in anaerobic digestion, resulting in increased process efficiency and stability.

- **Biodegradation and Composting:** Biodegradation, which is facilitated by microorganisms, is critical for breaking down organic matter in both natural and managed ecosystems. Composting is a controlled biodegradation process that converts organic waste into compost, which is a valuable soil additive. Bacteria, fungi, and actinomycetes decompose organic material, generating heat that sterilizes the compost by killing pathogens and weed seeds. Microbial inoculants have improved composting efficiency and quality, increasing nutrient availability and soil health (Malik and Goyal, 2022).
- **Phytoremediation:** Phytoremediation is the use of plants and microorganisms to remove, degrade, or stabilize environmental contaminants. Microorganisms in the rhizosphere, or soil region influenced by plant roots, play an important role in increasing phytoremediation efficiency. Certain rhizobacteria, for example, can promote plant growth while also increasing heavy metal uptake, making it easier to remove them from contaminated sites.

Recent advances have centered on genetic engineering to create plants with improved phytoremediation abilities. Transgenic plants that express bacterial genes for heavy metal transport and sequestration show promise for soil remediation (Tan et al., 2023). Furthermore, understanding the interactions between plants and their microbial partners has resulted in the creation of tailored microbial consortia that improve phytoremediation outcomes.

- **Bioenergy:** Microbial processes are used to generate bioenergy, providing a sustainable alternative to fossil fuels. In addition to anaerobic digestion for biogas production, microbial fuel cells (MFCs) convert organic matter into electricity using electrogenic bacteria's metabolic activities. MFC technology has advanced, including the development of novel electrode materials and microbial consortia, increasing energy conversion efficiency and scalability (Tan et al., 2023).
- **Carbon Sequestration**: Microorganisms play an important role in carbon sequestration, which reduces climate change by capturing and storing atmospheric carbon dioxide (CO_2). Soil microorganisms help form and stabilize soil organic matter, which is a significant carbon sink. Additionally, marine microorganisms such as phytoplankton and cyanobacteria drive the biological carbon pump, which sequesters carbon in ocean sediments. Enhancing microbial carbon sequestration through practices like biochar application and conservation tillage can help mitigate climate change (Maier et al., 2009; Tan et al., 2023).
- **Environmental Monitoring:** Microorganisms are increasingly being used as bioindicators in environmental monitoring, providing information about ecosystem health and quality. For example, the presence and abundance of certain microbial taxa can indicate water quality and pollution levels. Advances in molecular biology techniques, such as quantitative PCR (qPCR) and next-generation sequencing (NGS), have increased the sensitivity and accuracy of microbial monitoring, allowing for early detection of environmental changes and pollutants (Maier et al., 2009).

7.4. PHARMACEUTICAL MICROBIOLOGY

Pharmaceutical microbiology uses microorganisms to develop, produce, and test drugs. This discipline has a wide range of applications, including the discovery of new antibiotics, the production of vaccines, biopharmaceuticals, and enzymes, and ensuring the safety and efficacy of pharmaceutical products. Significant advances and technologies in pharmaceutical microbiology have transformed medicine, helping to improve patient outcomes (Malik et al., 2022).

7.4.1. Discovery of Antibiotics

The discovery of antibiotics was one of microbiology's most significant contributions to medicine. The antibiotic era began in 1928 when Alexander Fleming discovered penicillin, which is produced by the fungus *Penicillium notatum*. This breakthrough paved the way for the discovery of numerous other antibiotics derived from various bacterial and fungal species, such as streptomycin, tetracycline, and erythromycin (Baltz et al., 2010). The development of high-throughput screening and metagenomics has accelerated the discovery of new antibiotics from previously uncultivable microorganisms, addressing the growing problem of antibiotic resistance.

7.4.2. Production of Vaccines

Microorganisms play a critical role in vaccine development and production. Traditional vaccine production methods frequently employ attenuated or inactivated pathogens, as seen in polio and influenza vaccines. Advances in genetic engineering have enabled the creation of recombinant vaccines, which use GMMs to express specific antigens. For example, the Hepatitis B vaccine is made from recombinant yeast cells that express the Hepatitis B surface antigen (Gilmore and Denyer, 2023). Furthermore, the recent development of mRNA vaccines, such as the Pfizer-BioNTech and Moderna COVID-19 vaccines, represents a significant step forward, as they use lipid nanoparticles to deliver mRNA that encodes the viral spike protein.

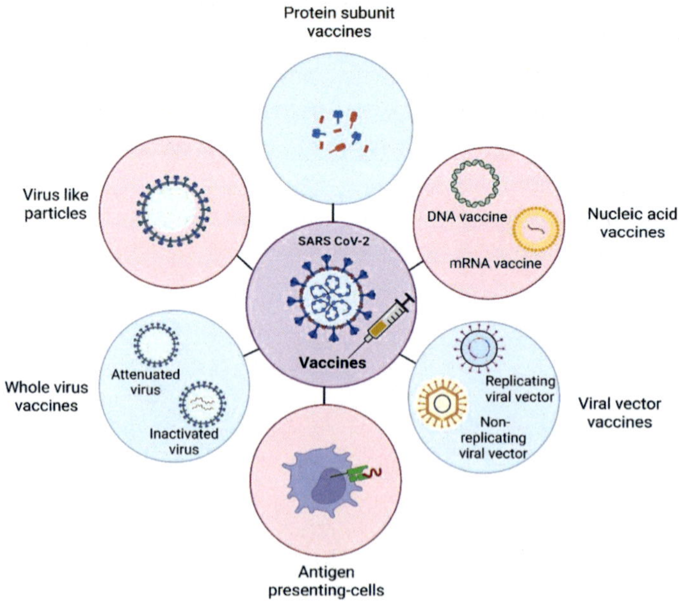

Figure 7.3. *An overview of COVID-19 vaccine platforms.*

Source: Chavda VP, Vihol DR, Solanki HK, Apostolopoulos V. (2022). The vaccine world of COVID-19: India's CONTRIBUTION. Vaccines, 10(11), 1943. https://doi.org/10.3390/vaccines10111943.

7.4.3. Biopharmaceuticals and Enzyme Production

Microbial expression systems are often used to produce biopharmaceuticals such as monoclonal antibodies, hormones, and therapeutic proteins. Because of their well-characterized genetics and high yield, *Escherichia coli* and yeast (*Saccharomyces cerevisiae*) are frequently used as hosts for recombinant protein production (Malik et al., 2022). For example, insulin, a critical hormone for diabetes management, is produced in *E. coli* using recombinant DNA technology. Furthermore, microbial enzymes are widely used in the pharmaceutical industry for drug synthesis and biotransformation. Microorganisms produce enzymes such as lipases, proteases, and amylases, which are then used to make various pharmaceuticals (Gilmore and Denyer, 2023).

7.4.4. Microbial Fermentation in Pharmaceutical Production

Microbial fermentation is a key component of pharmaceutical production, allowing for large-scale synthesis of various drugs and bioactive compounds. Antibiotics, vitamins, amino acids, and other secondary metabolites are produced through fermentation processes that involve bacteria, fungi, and yeast. Penicillin fermentation, for example, involves a submerged culture of *Penicillium chrysogenum* under controlled conditions to maximize yield (Baltz et al., 2010). Advances in fermentation technology, such as fermentation condition optimization and metabolic engineering, have significantly increased drug production efficiency and scalability.

7.4.5. Advanced Drug Delivery Systems

Microorganisms are being investigated as vehicles for advanced drug delivery. Bacterial vectors, such as attenuated Salmonella and Listeria, have been designed to deliver therapeutic agents directly to tumors and inflamed tissues. These vectors can be genetically modified to express therapeutic proteins or carry drug-loaded nanoparticles, allowing for drug release that is targeted and controlled. Furthermore, microbial-derived exosomes and outer membrane vesicles are being studied for their ability to deliver nucleic acids and small molecules for therapeutic purposes (Gilmore and Denyer, 2023).

7.4.6. Bioprospecting and Novel Drug Discovery

Bioprospecting entails investigating various environments, such as soil, marine ecosystems, and extreme habitats, in order to discover novel microorganisms with potential pharmaceutical applications. The search for new bioactive compounds from extremophiles, such as thermophiles and halophiles, has resulted in the discovery of novel enzymes and metabolites with therapeutic applications (Baltz et al., 2010; Malik et al., 2022). Advances in metagenomics and synthetic biology have made it possible to identify and exploit novel biosynthetic gene clusters, allowing for the development of new drugs and bioactive compounds.

7.4.7. Quality Control and Safety Testing

Ensuring the safety and efficacy of pharmaceutical products is critical, and microbiology plays an important role in quality control and risk assessment. Microbiological assays detect and quantify microbial contamination in pharmaceutical products, ensuring compliance with regulatory requirements. In pharmaceutical manufacturing, quality control includes sterility testing, microbial limit testing, and endotoxin testing. Rapid microbiological methods (RMMs) based on technologies such as PCR, flow cytometry, and MALDI-TOF MS have improved microbial testing speed and accuracy (Malik et al., 2022).

7.4.8. Personalized Medicine and Microbiome Research

The human microbiome has emerged as an important area of study, with implications for personalized medicine. The composition of a person's microbiome can affect drug metabolism, efficacy, and toxicity. Microbiome research is shedding light on the role of gut microbiota in health and disease, potentially leading to the development of microbiome-based therapies and probiotics. Fecal microbiota transplantation (FMT) is an example of a microbiome-based therapy used to treat recurrent *Clostridium difficile* infections, demonstrating the therapeutic potential of microbiome manipulation (Murray et al., 2015; Gilmore and Denyer, 2023).

7.5. FOOD MICROBIOLOGY

Microbiology has a wide-ranging impact on food production, safety, and quality. Microbiology has a wide range of applications in this sector, including food fermentation, biopreservation, food safety and spoilage control, and the development of functional foods and probiotics. Significant advancements and technologies in food microbiology have improved our ability to produce safe, nutritious, and high-quality food products that meet both consumer and regulatory needs (Baltz et al., 2010).

7.5.1. Fermentation Processes

Fermentation is one of the oldest applications of microbiology in the food industry, used to produce a wide range of foods and beverages. Biochemical processes use microorganisms like bacteria, yeasts, and molds to convert raw ingredients into fermented products. *Saccharomyces cerevisiae*, for example, ferments sugars into ethanol and carbon dioxide in the production of bread, beer, and wine (Baltz et al., 2010). Similarly, LAB such as Lactobacillus and Streptococcus species are required for the production of dairy products like yogurt and cheese, as well as fermented vegetables like sauerkraut and kimchi.

Advances in fermentation technology have resulted in the development of controlled fermentation processes, which ensure consistent product quality and safety. The use of starter cultures, which are microorganism strains with desirable fermentation characteristics, has increased the reliability and efficiency of fermentation processes (Da Silva et al., 2018). Furthermore, advances in molecular biology and genomics have allowed for genetic modification of microorganisms to improve fermentation performance and product characteristics.

7.5.2. Functional Foods and Probiotics

Functional foods and probiotics are important applications of microbiology in the food industry, focusing on increasing the health benefits of food products. Probiotics are living microorganisms that provide health benefits to the host when consumed in sufficient quantities. *Lactobacillus* and *Bifidobacterium* are two common probiotic strains found in dairy products, dietary supplements, and functional foods (Rao, 2016).

Advances in probiotic research have resulted in the development of products that address specific health issues, such as gastrointestinal disorders, immune modulation, and metabolic syndrome. Encapsulation technology has increased the viability and stability of probiotics during processing and storage, ensuring their effectiveness when consumed. Furthermore, the study of the gut microbiome has shed light on the mechanisms by which probiotics exert their beneficial effects, guiding the development of next-generation probiotic formulations (Jagadeesan et al., 2019).

7.5.3. Biopreservation

Biopreservation is the use of natural or controlled microbiota and their antimicrobial compounds to increase the shelf life and safety of food products. This strategy takes advantage of the antimicrobial properties of specific microorganisms and their metabolites, such as organic acids, bacteriocins, and hydrogen peroxide. LAB, for example, produce lactic acid and bacteriocins, which prevent the growth of spoilage and pathogenic microorganisms (Jagadeesan et al., 2019).

Nisin, a bacteriocin produced by *Lactococcus lactis*, is a widely used food preservative. It works against a wide variety of Gram-positive bacteria, including

Listeria monocytogenes, a major foodborne pathogen. Biopreservatives have recently been incorporated into food packaging and coatings, adding an extra layer of protection against microbial contamination (Murray et al., 2015).

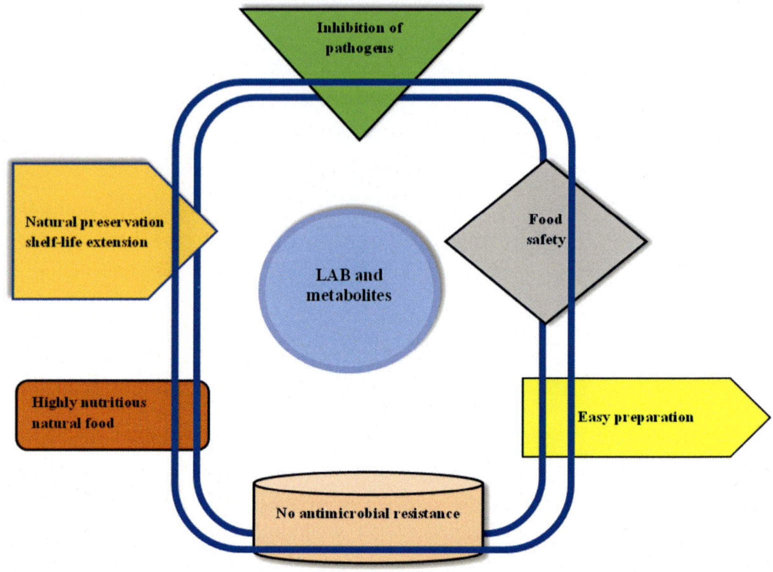

Figure 7.4. *Major advances in the field of biopreservation thanks to omics approaches.*

Source: Cortés-Sánchez ADJ, Jaramillo-Flores ME, Díaz-Ramírez M, Espinosa-Chaurand LD, Torres-Ochoa E. (2024). Biopreservation and the safety of fish and fish products, the case of lactic acid bacteria: A basic perspective. Fishes, 9(8), 303. https://doi.org/10.3390/fishes9080303.

7.5.4. Biotechnology and Genetic Engineering

Biotechnology and genetic engineering have created new opportunities for the use of microbiology in the food industry. Microorganisms' genetic modification can improve their fermentation capabilities, stress tolerance, and metabolite production. For example, genetically modified yeasts are used to produce bioethanol more efficiently, whereas recombinant LAB can produce functional ingredients like vitamins and antimicrobial peptides (Jagadeesan et al., 2019).

Synthetic biology is a growing field that involves designing and building new biological parts, devices, and systems for specific applications. Synthetic biology can be used in the food industry to develop custom microbial strains with optimized metabolic pathways for producing new flavors, colors, and functional ingredients (Baltz et al., 2010). CRISPR-Cas9 technology has further revolutionized genetic engineering by allowing for precise editing of microbial genomes, making it easier to develop tailored microbial strains for a variety of food applications.

7.5.5. Food Safety and Spoilage Control

To avoid foodborne illnesses and spoilage, it is critical to ensure microbiological food safety. Salmonella, *Escherichia coli* O157, *Listeria monocytogenes*, and Campylobacter are among the foodborne pathogens that pose significant public health risks (Da Silva et al., 2018). Traditional microbiological methods for detecting these pathogens rely on culture-based techniques, which are time-consuming and labor-intensive.

Recent advances in rapid detection techniques have greatly improved food safety monitoring. Polymerase chain reaction (PCR) and qPCR are molecular techniques that enable the rapid and specific detection of foodborne pathogens. NGS and metagenomics offer comprehensive insights into the microbial communities found in food, allowing the identification of both culturable and non-culturable microorganisms (Baltz et al., 2010; Jagadeesan et al., 2019). Immunoassays, such as enzyme-linked immunosorbent assays (ELISA), provide another fast and sensitive way to detect specific pathogens or their toxins in food samples.

7.6. BIOREMEDIATION

Bioremediation is a process that uses microbes, plants, or microbial or plant enzymes to clean up and restore polluted ecosystems. It is a natural process that uses biological organisms to degrade or change toxins into less toxic forms, thereby minimizing the detrimental effects of pollutants (Megharaj et al., 2011). This technology is especially beneficial because it is environmentally friendly and cost-effective, offering an alternative to more invasive and costly restoration procedures.

7.6.1. Microbial Processes in Bioremediation

The foundation of bioremediation is microbial activities, which use the metabolic pathways of bacteria, fungi, and other microorganisms to degrade contaminants. Through both aerobic and anaerobic mechanisms, these microbes are able to break down organic contaminants. Microbes utilize oxygen in aerobic bioremediation to oxidize pollutants and transform them into carbon dioxide, water, and biomass (Saeed et al., 2022). This works especially well for contaminants like hydrocarbons from oil spills. On the other hand, anaerobic bioremediation takes place in oxygen-starved settings where bacteria degrade pollutants using other electron acceptors such as nitrate, sulfate, or iron. Both heavy metals and chlorinated chemicals can be treated using this technique.

Microbial communities are frequently improved through bioaugmentation, in which specific strains recognized for their breakdown abilities are added to the contaminated location. In addition, biostimulation techniques are used to improve circumstances for indigenous microbial activity by adding nutrients or changing environmental parameters (Megharaj et al., 2011).

7.6.2. Techniques and Technologies in Bioremediation

Bioremediation employs a variety of methodologies and technologies, each adapted to the type and extent of contamination, as well as the environmental circumstances. These approaches are basically divided into two categories: In situ and ex situ techniques (Saeed et al., 2022). In situ approaches entail treating contaminated material on-site. This includes:

- **Bioventing**: Improves the activity of native microorganisms by creating appropriate air and nutrient conditions.
- **Phytoremediation**: Uses plants to absorb, collect, and break down contaminants in soil and water.
- **Biosparging**: Involves injecting pressurized air beneath the water table to boost oxygen content and stimulate microbial activity.

Ex situ procedures include the removal of contaminated material for treatment elsewhere. This includes:

- **Biopiles**: Contaminated soil is dug and deposited in piles under regulated conditions to promote microbial activity.
- **Windrow Composting:** Involves aeration and occasional rotating of organic debris to promote microbial decomposition.
- **Bioreactors**: Utilize designed systems with conditions optimized for microbial breakdown of pollutants (Saeed et al., 2022).

7.6.3. Monitoring and Evaluation

To gauge the success and advancement of the remediation process, monitoring and evaluation serve as vital elements of bioremediation programs. These operations entail the routine collection and analysis of environmental samples, including soil, water, and air to evaluate pollutant levels and microbial activity (Lederberg and Relman, 2009).

Key techniques used in monitoring include:

- **Chemical Analysis**: Determining the amounts of contaminants by techniques such as high-performance liquid chromatography (HPLC) and gas chromatography-mass spectrometry (GC-MS).

- **Microbial Analysis**: This is the process of evaluating microbial populations and activity using molecular methods like DNA sequencing and PCR.
- **Environmental Parameters**: To guarantee ideal circumstances for microbial activity, keep an eye on pH, temperature, moisture content, and nutrient levels.
- Sufficient monitoring ensures that the bioremediation process is proceeding as planned and permits prompt modifications to improve performance (Lederberg and Relman, 2009).

7.6.4. Challenges and Future Directions of Bioremediation

Bioremediation presents a number of difficulties despite its benefits. The fluctuating nature of site conditions, which can impact pollutant bioavailability and microbial activity, is one major concern. Bioremediation processes may be hindered by the uneven dispersion of pollutants and environmental variables including pH, temperature, and moisture content. Furthermore, some contaminants are resistant to microbial breakdown, which calls for the development of more sophisticated bioremediation techniques (Saeed et al., 2022).

The primary objective of bioremediation research in the future is to use cutting-edge technologies to overcome these obstacles. Synthetic biology and genetic engineering developments could potentially produce microbes that are more resilient to environmental stressors and have improved degrading capacities. Nanoparticles are being used in bioremediation to enhance the efficacy and delivery of microbial consortia, which is another new application of nanotechnology in biotechnology.

Additionally, the use of omics technologies—genomics, proteomics, and metabolomics—to learn more about microbial populations and their interactions with pollutants is becoming more and more popular. Bioremediation techniques that are more focused and effective may result from this all-encompassing approach. Last but not least, combining bioremediation with other environmentally friendly activities like phytomining and the generation of bioenergy may offer further financial and environmental advantages, encouraging the wider use of bioremediation methods (Megharaj et al., 2011).

7.7. BIOMINING

Biomining is a novel biotechnological method in which microorganisms extract rich metals from ores and other solid materials. Unlike traditional mining methods, which rely on chemical and physical means, biomining uses the natural metabolic activities of specific microorganisms to solubilize metals, making them easier to recover. This environmentally benign method is especially useful for processing low-grade ores and mining waste, where traditional processes may be economically prohibitive. Biomining provides a sustainable alternative to traditional mining techniques, potentially lowering environmental impact and operational costs (Johnson, 2014).

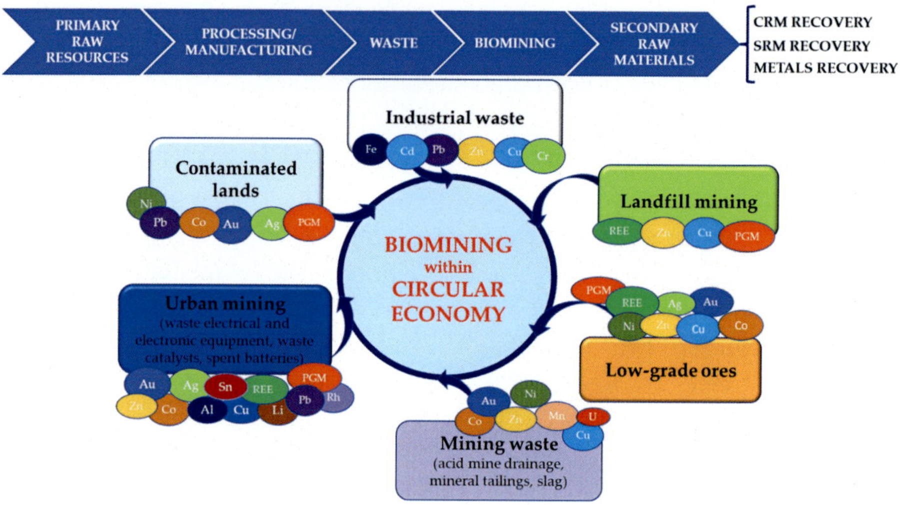

Figure 7.5. *Graphic abstract of biomining.*

Source: Cozma P, Be□ianu C, Hlihor R-M, Simion IM, Gavrilescu M. (2024). Bio-recovery of metals through biomining within circularity-based solutions. Processes, 12(9), 1793. https://doi.org/10.3390/pr12091793.

7.7.1. Principles of Biomining

Biomining principles are based on microbial metabolism, particularly certain microorganisms' oxidative and reductive metabolic pathways. These microbes, such as bacteria and archaea, obtain energy by the oxidation of inorganic substances, a process known as chemolithotrophy. Biomining is primarily concerned with the oxidation of sulfide minerals. Microorganisms like *Acidithiobacillus ferrooxidans* convert ferrous iron (Fe^{2+}) to ferric iron (Fe^{3+}) and sulfide minerals to sulfate (SO_4^{2-}), allowing for the release of metal ions like copper, gold, and uranium into solution. pH, temperature, and the presence of particular ions all have an impact on microbial activity and metal recovery rates throughout the bioleaching process (Johnson, 2014).

7.7.2. Microorganisms Used in Biomining

Several microorganisms are important in biomining because of their metabolic capacities and adaptation to harsh settings. The most prominent are:

- ***Acidithiobacillus ferrooxidans*:** This acidophilic bacterium is commonly utilized in biomining because of its capacity to oxidize ferrous iron and sulfide minerals.
- ***Acidithiobacillus thiooxidans*:** This bacterium, known for oxidizing sulfur compounds, works alongside *A. ferrooxidans* in sulfide mineral bioleaching.
- ***Leptospirillum ferrooxidans*:** Another iron-oxidizing bacterium commonly discovered in biomining consortiums.
- **Sulfobacillus thermosulfidooxidans:** This is a thermophilic bacterium that grows at higher temperatures, making it suited for treating specific ores.
- ***Ferroplasma acidarmanus:*** An archaeon that oxidizes iron and thrives in highly acidic situations (Roberto and Schippers, 2022).

These microorganisms not only enhance metal recovery but also exhibit resilience in the harsh conditions of biomining operations, such as low pH and high metal concentrations.

7.7.3. Biomining Processes and Techniques

Biomining comprises a variety of processes and procedures, which are broadly classified as bioleaching and bio-oxidation.

1. **Bioleaching:** This approach is the microbial conversion of insoluble metal sulfides into soluble metal sulfates, which aids in the extraction of metals such as copper, nickel, and zinc. The heap leaching process is widely employed, in which crushed ore is heaped and watered with a leaching solution containing microorganisms.
2. **Biooxidation**: Specifically used for gold recovery, bio-oxidation involves the microbial destruction of the mineral matrix enclosing gold, revealing the metal for subsequent chemical extraction. This technique is commonly used for refractory ores that are resistant to traditional cyanidation.
3. **In Situ biomining**: This process includes applying microbial solutions directly to ore bodies that are still on the earth. This technology is especially effective for ores that are difficult to obtain or would be prohibitively expensive to mine using traditional techniques.

4. **Stirred tank bioreactors**: This approach agitates and aerates a slurry of ore and microorganisms in enormous tanks to optimize microbial activity and metal extraction (Roberto and Schippers, 2022).

Each approach has advantages and disadvantages, with the method chosen based on ore type, economic considerations, and environmental impact.

7.7.4. Environmental and Economic Implications

There are significant financial and environmental implications to biomining. Biomining is good for the environment since it uses less toxic chemicals and leaves less environmental footprints from mining activities. Compared to traditional mining processes, the approach produces less harmful waste and greenhouse gases. Additionally, by detoxifying mining waste, biomining can aid in the cleanup of contaminated locations (Johnson, 2014).

Economically, biomining provides low-cost methods of removing metals from low-grade ores and mining debris, which are frequently regarded as waste in conventional mining. Recovering valuable metals from these sources increases resource efficiency and prolongs mining operations. Additionally, biomining lowers operating expenses by consuming less energy throughout the ore processing process (Roberto and Schippers, 2022).

Nevertheless, there are difficulties in putting biomining into practice. The effectiveness of metal recovery can vary based on microbial performance and ambient circumstances, and the process may take longer than traditional procedures. Despite these obstacles, biomining remains a viable strategy for sustainable mining in the future as research and technology developments keep enhancing its viability and efficiency (Johnson, 2014).

7.8. BIOFUELS

Biofuels, derived from biological materials, are considered a sustainable alternative to fossil fuels. The microbial production of biofuels harnesses the metabolic capabilities of microorganisms to convert biomass into energy-dense compounds (Wijffels and Barbosa, 2010). This approach leverages the diversity and efficiency of microbial metabolisms to produce biofuels such as ethanol, biodiesel, and biogas, offering a promising solution to global energy challenges.

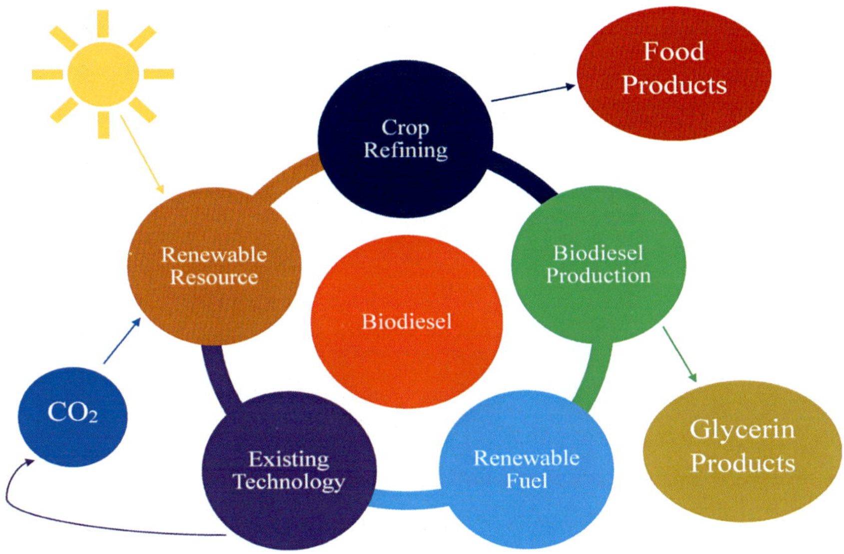

Figure 7.6. *Flow diagram of the biodiesel making process.*

Source: Zulqarnain, Ayoub M, Yusoff MHM, Nazir MH, Zahid I, Ameen M, Sher F, Floresyona D, Budi Nursanto E. (2021). A comprehensive review on oil extraction and biodiesel production technologies. Sustainability, 13(2), 788. https://doi.org/10.3390/su13020788.

7.8.1. Microorganisms Involved in Biofuels Production

Various microorganisms are pivotal in the production of biofuels due to their ability to degrade complex biomolecules into simpler forms that can be readily converted into energy. Key microorganisms include:

1. **Bacteria** like *Clostridium acetobutylicum* are utilized to produce butanol and other solvents. *Zymomonas mobilis* is also known for its high ethanol output and tolerance of inhibitory chemicals present in lignocellulosic biomass.
2. **Yeasts**: *Saccharomyces cerevisiae* is the most commonly utilized yeast in bioethanol production due to its durability and efficiency at digesting carbohydrates. Another yeast, *Pichia pastoris*, has shown promise for making biofuels from non-sugar substrates.
3. **Algae**: Microalgae, such as Chlorella and Nannochloropsis, are used in biodiesel production due to their high lipid content and rapid growth rates. They may be cultivated in a variety of conditions, including wastewater, which increases their sustainability (Johnson, 2014).

7.8.2. Genetic Engineering and Optimization

Microorganisms can be optimized through genetic engineering to produce more biofuel. Among the strategies are:

- **Metabolic Pathway Engineering:** Changing microorganisms' metabolic pathways to produce more of the desired biofuels. For instance, expressing heterologours pathways in _*Escherichia coli* to produce isobutanol and other alcohols has been demonstrated to significantly increase yield.
- **Tolerance Enhancement:** Creating strains with increased tolerance to biofuel and inhibitor concentrations found in biomass hydrolysates is known as tolerance enhancement. Methods include upregulating stress response pathways or introducing genes that give resistance.
- **Synthetic Biology:** Using techniques from synthetic biology to create new pathways and maximize gene expression. This includes creating artificial regulatory circuits to regulate metabolic fluxes and precisely altering genomes using CRISPR/Cas9 (Nielsen and Keasling, 2016).

7.8.3. Biofuels Production Process

The production of biofuels involves several stages:

- **Feedstock Preparation:** Complex polymers are broken down into fermentable sugars by pre-treating biomass, such as algae, agricultural wastes, or special energy crops. Among the techniques include enzymatic, chemical, and mechanical processes.
- **Fermentation:** The sugars are converted into biofuels by microorganisms. Depending on the required scale and efficiency, this process can be carried out in batch, fed-batch, or continuous modes. In order to maximize yields, it is essential to optimize fermentation conditions, including pH, temperature, and nutrient availability.
- **Recovery and Purification:** Distillation, extraction, or other purification methods are used to remove the biofuels from the fermentation broth. The goals of developments in integrated bioprocessing systems and membrane technologies are to lower energy usage and increase recovery efficiency (Nielsen and Keasling, 2016).

7.8.4. Environmental, Economic, and Regulatory Considerations

The adoption of microbial biofuels involves several considerations:

- **Environmental Impact**: Microbial biofuels have the potential to dramatically cut greenhouse gas emissions when compared to fossil fuels. The use of waste biomass and algae cultivation in wastewater can help to reduce pollution and encourage sustainable practices.
- **Economic Viability**: The cost-effectiveness of biofuels is determined by feedstock availability, production efficiency, and market dynamics. Advances in genetic engineering and process optimization promise to lower production costs and improve competitiveness with traditional fuels.
- **Regulatory Framework**: To assure safety, sustainability, and environmental protection, microbial biofuels are developed and commercialized following regulatory standards. Policies that promote renewable energy and incentivize biofuel production are crucial to fostering industry growth (Wijffels and Barbosa, 2010).

7.9. BIOPHARMACEUTICALS

Biopharmaceuticals are medicinal compounds generated from biological sources, typically proteins, nucleic acids, or living cells, and are used to prevent, diagnose, and treat diseases. Biopharmaceuticals, as opposed to typical chemically synthesized medications, are generated using biotechnology processes that rely on live creatures such as bacteria, yeast, and mammals. These items include monoclonal antibodies, vaccinations, gene treatments, and recombinant proteins (Hewitt, 2004). Their creation represents a big step forward in medicine, providing focused therapy options with more specificity and fewer side effects.

7.9.1. Microbial Production System

A key component of the biopharmaceutical manufacturing process is the microbial production system, which uses microorganisms like yeast and bacteria to create therapeutic proteins and other biological products. Because of its capacity to create recombinant proteins, well-understood genetics, and rapid growth, *Escherichia coli*, or *E. coli*, is one of the most widely utilized bacterial systems. One benefit of using yeasts such as *Saccharomyces cerevisiae* is that they can provide post-translational modifications, which are necessary for many eukaryotic proteins to function. By introducing the desired gene into these genetically modified microbial hosts, fermentation processes can produce significant amounts of the targeted biopharmaceuticals. The industry favors microbial systems due to their high yield, cost-effectiveness, and scalability (Johnson, 2014).

7.9.2. Types of Biopharmaceuticals

Biopharmaceuticals are a broad category of products that include:

1. **Monoclonal Antibodies**: Widely utilized in the treatment of infectious diseases, autoimmune disorders, and malignancies, these engineered antibodies are made to target particular antigens. Examples are infliximab for rheumatoid arthritis and trastuzumab for breast cancer.
2. **Vaccines:** Immunization against infectious diseases via biological preparations. Recombinant vaccines, such as the HPV vaccination, are made using microbial systems thanks to modern breakthroughs.
3. **Recombinant Proteins:** Therapeutic proteins made possible by recombinant DNA technology, such as erythropoietin for the treatment of anemia and insulin for the management of diabetes.
4. **Gene Therapies:** To treat or prevent disease, genetic material is delivered into cells. Viral vectors and CRISPR-based systems are among the methods.
5. **Cell Therapies:** Using live cells to cure illnesses; an example of this is CAR-T cell treatment for specific cancer types (Hewitt, 2004).

7.9.3. Bioprocessing and Manufacturing

In the production of biopharmaceuticals, bioprocessing entails several crucial steps:

- **Upstream Processing:** This comprises fermentation, media optimization, and cell line creation. The objective is to use microbial or cell culture systems to maximize the synthesis of the target chemical.
- **Downstream Processing:** This step, which comes after fermentation, concentrates on recovering and purifying the biopharmaceutical product. To guarantee the high purity and activity of the finished product, methods like centrifugation, chromatography, ultrafiltration, and filtration are used.
- **Formulation and Fill-Finish:** To guarantee stability, effectiveness, and safe distribution, the purified product is prepared with excipients. In the fill-finish process, the product is sterilely filled into syringes or vials so that it is ready for distribution.
- **Quality Control and Assurance:** Making sure, via thorough testing and validation procedures, that the product satisfies strict regulatory criteria for safety, efficacy, and consistency (Hewitt, 2004).

7.9.4. Regulatory and Ethical Considerations

Biopharmaceuticals have strict regulatory requirements, reflecting the complexity and possible dangers connected with biological products. Regulatory bodies such as the United States Food and Drug Administration (FDA) and the European Medicines Agency (EMA) impose detailed criteria for clinical trials, manufacturing processes, quality control, and post-marketing surveillance. Important regulatory factors include:

1. **Safety and Efficacy:** Showing that the biopharmaceutical is both safe and effective through preclinical investigations and multi-phase clinical trials.
2. **Good Manufacturing Practice (GMP):** Following GMP guidelines is crucial to ensuring product quality and uniformity. This includes upholding high standards in production facilities, employee training, and documentation.
3. **Post-Market Surveillance:** Continuous monitoring of the biopharmaceutical market performance to identify any undesirable effects or long-term concerns.

Ethical considerations in biopharmaceuticals include obtaining patient consent and assuring clinical trial transparency, addressing therapy access and affordability, and upholding ethical norms in genetic alteration and stem cell research. The promise of biopharmaceuticals comes with the duty to carefully and rigorously traverse these regulatory and ethical contexts (Wijffels and Barbosa, 2010).

Therefore, biopharmaceuticals mark an important turning point in medical therapy, leveraging the capabilities of biotechnology and microbiology to create sophisticated medicines. Their manufacture requires advanced microbial systems and bioprocessing processes, which are supported by strict regulatory and ethical frameworks to ensure the greatest levels of patient safety and product efficacy.

7.10. SYNTHETIC BIOLOGY

In the multidisciplinary field of synthetic biology, biological and engineering principles are combined to create new biological components, devices, and systems, or to repurpose already-existing biological systems. The goal of this quickly developing profession is to get a deeper understanding of biological processes and use them for creative applications in a variety of disciplines, such as environmental science, agriculture, and medicine.

7.10.1. Foundational Techniques and Tools

Synthetic biology is based on a set of sophisticated methods and instruments that allow for the accurate modification of genetic material. Technologies for DNA synthesis and sequencing are important examples of this, as they have become much more efficient and affordable. With previously unheard-of precision, precise alterations to DNA sequences can be made using the groundbreaking genome-editing technology CRISPR-Cas9 (Mao et al., 2021). Furthermore, the creation of standardized biological components, sometimes known as "BioBricks," makes the construction of intricate genetic structures easier.

Computational modeling and bioinformatics also play crucial roles in synthetic biology. These tools assist in the design of genetic circuits and the prediction of their behavior in living organisms. Techniques such as Gibson assembly and Golden Gate cloning are essential for the seamless assembly of multiple DNA fragments, while synthetic gene networks and synthetic promoters enable precise control over gene expression (Lederberg and Relman, 2009).

7.10.2. Microbial Applications

Synthetic biology has had a significant impact on microbial applications since it allows microbes to be engineered for a variety of functions. Engineered microorganisms, such as *Escherichia coli* and yeast, have been created to accomplish certain jobs that traditional methods cannot efficiently complete. By altering microbial metabolic pathways, synthetic biology has made it easier to produce biofuels, medicines, and specialty compounds. Studies have shown that *E. coli* may be genetically manipulated to produce high-value chemicals such as artemisinin, a crucial antimalarial medicine. In addition, synthetic biology has boosted bioremediation efforts by engineering microbes to break down environmental contaminants or neutralize dangerous compounds (Mao et al., 2021). These applications take advantage of synthetic biology's ability to optimize metabolic pathways and improve microbial performance, solving industrial and environmental concerns.

7.10.3. Cellular and Organismal Engineering

Cellular and organismal engineering are critical components of synthetic biology, which focuses on reprogramming cells and organisms to perform new roles. Synthetic biology allows for the generation of minimal cells with streamlined genomes that are easier to engineer (Gibson et al., 2010). These tiny cells serve as platforms for investigating fundamental biological processes and creating new biological activities. Furthermore, cellular reprogramming techniques enable the conversion of normal cells into specialized types with therapeutic properties, such as modified T-cells for cancer immunotherapy. The development of synthetic creatures with artificially generated genomes offers a new frontier in biological study, bringing insights into life's principles as well as instruments for fresh applications (Johnson, 2014).

7.10.4. Application and Innovation

Synthetic biology is yielding numerous and revolutionary applications and discoveries. In biotechnology, complex compounds, such as medications and materials, are produced using GMMs. For instance, by creating bacteria that can manufacture antibiotics and therapeutic proteins more quickly than with conventional techniques, synthetic biology has proven useful in the development of novel drugs. Furthermore, engineered organisms are being produced for functions like carbon sequestration and garbage treatment, adding to eco-friendly technology. This shows promise for the study of environmental sustainability and synthetic biology. It is anticipated that additional advancements in synthetic biology will result in the creation of sophisticated biological systems and applications with hitherto unheard-of capabilities, as a result of its integration with other fields like nanotechnology and artificial intelligence (Mao et al., 2021).

CHAPTER 8

MICROBIAL PATHOGENICITY

Contents

Microbial pathogenicity is a fundamental topic in microbiology that covers the mechanisms by which microorganisms cause disease in host organisms. Virulence factors, host-pathogen interactions, and molecular pathways that facilitate microbial invasion and colonization are central concepts in this field. Pathogenic microbes, such as bacteria, viruses, fungi, and parasites, have a variety of virulence factors, including toxins, enzymes, and surface proteins, which allow them to penetrate host defenses, evade the immune system, and cause cellular damage (Maier et al., 2009; Murray et al., 2015).

The study of microbial pathogenicity delves into the genetic and biochemical foundations of these virulent mechanisms. For example, bacterial pathogens frequently employ specialized secretion systems, such as the Type III secretion system, to inject effector proteins directly into host cells, manipulating host cellular processes to their advantage (Meena et al., 2019). Furthermore, pathogens' ability to form biofilms is critical for their survival and persistence, complicating treatment and eradication efforts.

Host-pathogen interactions are dynamic and complex, with microbial virulence strategies competing with host immune responses continuously. Understanding these interactions at the molecular level is critical for developing effective therapeutic interventions and prevention strategies for infectious diseases (Madigan et al., 2018).

8.1. HOST-PATHOGEN INTERACTIONS

Understanding microbial pathogenicity requires an understanding of host-pathogen interactions, which involve a dynamic interplay between pathogens' invasive strategies and the host's defensive mechanisms. This intricate relationship has a significant impact on infection outcomes, determining disease severity and progression. Several key aspects of host-pathogen interactions are critical for understanding the mechanisms underlying microbial pathogenicity (Madigan et al., 2018).

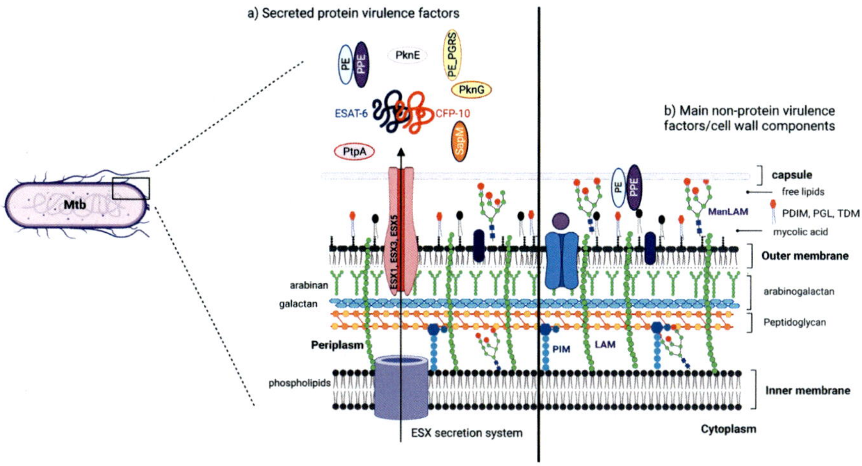

Figure 8.1. *Virulence factors of Mycobacterium tuberculosis.*

Source: Ramon-Luing LA, Palacios Y, Ruiz A, Téllez-Navarrete NA, Chavez-Galan L. (2023). Virulence factors of mycobacterium tuberculosis as modulators of cell death mechanisms. Pathogens, 12(6), 839. https://doi.org/10.3390/pathogens12060839.

8.1.1. Virulence Factors

Pathogens have a variety of virulence factors that promote infection and disease progression. Toxins, such as exotoxins and endotoxins, can cause tissue damage and disrupt normal cellular functions. Adhesion molecules such as pili, fimbriae, and adhesins allow pathogens to attach to host cells, which is an important step in colonization and infection (Meena et al., 2019). Invasion factors, including enzymes such as hyaluronidases and collagenases, help pathogens penetrate host tissues and spread to deeper tissues.

8.1.2. Pathogen Recognition and Immune Evasion

Pattern recognition receptors (PRRs), such as Toll-like receptors (TLRs), help host cells detect pathogen-associated molecular patterns (PAMPs). This recognition activates innate immune responses that seek to eliminate the invader (Tay et al., 2016). Pathogens, on the other hand, have evolved a variety of strategies for evading immune responses. Some bacteria, for example, can change their surface antigens through phase or antigenic variation, preventing the host immune system from recognizing them. Other mechanisms include the formation of capsules to prevent phagocytosis and the secretion of proteins that disrupt complement activation or antibody function (Maier et al., 2009).

8.1.3. Adhesion and Colonization

Successful colonization is an important step in microbial pathogenesis. Pathogens attach to host cells via specific adhesion molecules that recognize and bind to host cell receptors. This binding not only anchors the pathogen but also activates signaling pathways that aid in invasion. For example, uropathogenic *Escherichia coli* (UPEC) use pili to adhere to the urinary tract epithelium, which is an important step in causing urinary tract infections (Chatterjee and Raval, 2019).

8.1.4. Invasion Mechanisms

Once attached to host cells, pathogens can use a variety of strategies to invade and spread within the host. Some pathogens, such as Salmonella spp., use specialized secretion systems, such as the Type III secretion system, to deliver effector proteins to host cells. These proteins influence host cellular processes, facilitating pathogen uptake and intracellular survival. Other pathogens, such as Listeria monocytogenes, use host cell machinery to facilitate internalization and intracellular movement (Chatterjee and Raval, 2019).

8.1.5. Biofilm Formation

Biofilms are organized communities of microorganisms that live in an extracellular matrix that they create for themselves. Pathogens benefit from biofilm formation in a variety of ways, including increased antibiotic resistance and protection against host immune responses. *Pseudomonas aeruginosa*, a common opportunistic pathogen, forms biofilms that help it survive chronic infections like cystic fibrosis (Tay et al., 2016).

8.1.6. Host Immune Response

The host immune system combats infections using both innate and adaptive responses. Through physical barriers, phagocytic cells, and inflammatory responses, innate immunity serves as the first line of defense. Adaptive immunity, which includes T and B cells, produces a more specific and long-lasting response. Pathogens are constantly developing mechanisms to counteract these defenses, resulting in a coevolutionary arms race between host and pathogen (Madigan et al., 2018; Meena et al., 2019).

8.1.7. Quorum Sensing and Signal Transduction

Many pathogens use quorum sensing to regulate the expression of virulence factors based on population density. This cell-to-cell communication system enables pathogens to regulate their behavior in a coordinated manner, increasing their pathogenic potential (Chatterjee and Raval, 2019). Signal transduction pathways in both pathogens and host cells are critical for regulating these interactions and influencing infection outcomes.

8.1.8. Host Specificity and Tropism

Pathogen specificity for specific hosts or tissues is another important factor in host-pathogen interactions. The presence of specific receptors on host cells that pathogen adhesion molecules recognize is a common indicator of host specificity. Tropism, or pathogen preference for specific tissues, is influenced by factors such as tissue-specific receptor expression and local environmental conditions (Meena et al., 2019).

8.1.9. Pathogen Reservoirs and Transmission

Pathogens' pathogenicity is influenced by their ecology, which includes reservoirs and modes of transmission. Reservoirs can be either natural, such as soil and water, or biological, such as animal hosts. Understanding pathogen transmission dynamics is critical to developing effective control and prevention strategies (Maier et al., 2009; Murray et al., 2015).

8.2. VIRULENCE FACTORS

Microbial virulence factors are specialized molecules produced by pathogens that increase their ability to cause disease, and they play an important role in the complex process of microbial pathogenicity. These factors allow pathogens to enter host tissues, evade immune responses, obtain nutrients, and cause harm to the host (Meena et al., 2019). Understanding these virulence mechanisms is critical in microbiology because it explains how infections develop and persist and informs the development of therapeutic interventions.

8.2.1. Adhesion Factors

Adherence to host cells is an important first step in the infection process. Microbial adhesion factors, such as pili, fimbriae, and adhesins, help pathogens adhere to specific host tissues. For example, *Escherichia coli* fimbriae allow the bacteria to adhere to urinary tract epithelial cells, which is an important step in the pathogenesis of urinary tract infections (Tay et al., 2016). Adhesion not only anchors the pathogen but also activates host cell signaling pathways, promoting invasion and colonization. The specificity of these interactions often determines the pathogen's tissue tropism, which influences where infections occur within the host.

8.2.2. Invasion Factors

Invasion factors are virulence factors that allow pathogens to enter host tissues and spread throughout the body. Enzymes like hyaluronidases and collagenases degrade extracellular matrix components, allowing pathogens to penetrate physical barriers and enter deeper tissues. Another example is the Type III secretion system found in certain Gram-negative bacteria such as Salmonella and Shigella, which delivers effector proteins directly to host cells. These proteins influence host cell processes, promoting bacterial uptake, survival, and replication (Meena et al., 2019).

8.2.3. Toxins

Toxins are potent virulence factors that directly harm host tissues or disrupt normal cellular functions. Exotoxins and endotoxins are the two major types. Exotoxins, which bacteria produce, include neurotoxins, enterotoxins, and cytotoxins. *Clostridium botulinum*, for example, produces a neurotoxin known as botulinum toxin, which inhibits neurotransmitter release and causes paralysis. Cytotoxins, such as the alpha-toxin of *Staphylococcus aureus*, cause pores in host cell membranes, resulting in cell lysis and tissue damage (Chatterjee and Raval, 2019). Endotoxins, which include lipopolysaccharides (LPS), are components of Gram-negative bacteria's outer membrane. LPS, when released during bacterial lysis, causes a strong inflammatory response that can result in septic shock.

8.2.4. Immune Evasion Mechanisms

Pathogens have developed a variety of strategies to avoid the host immune system, ensuring their survival and persistence within the host. Antigenic variation is a common mechanism in which pathogens change their surface proteins to avoid recognition by the host's immune system. Neisseria gonorrhoeae, for example, frequently changes its pili proteins, preventing the host from mounting a strong immune response. Some bacteria, such as *Streptococcus pneumoniae*, produce polysaccharide capsules that inhibit phagocytosis, making it difficult for immune cells to engulf and destroy (Murray et al., 2015; Meena et al., 2019).

8.2.5. Biofilm Formation

Biofilms are microbial communities that form their own extracellular matrix. Biofilm formation is an important virulence factor because it increases pathogen resistance to antibiotics and protects them from the host immune system. Bacteria in a biofilm can communicate and coordinate their activities using quorum sensing, increasing their collective virulence (Maier et al., 2009). *Pseudomonas aeruginosa* is well-known for forming biofilms in chronic lung infections, particularly in cystic fibrosis patients, where the biofilm mode of growth contributes to pathogen persistence and treatment resistance.

8.2.6. Siderophores

Siderophores are molecules produced by bacteria that scavenge iron from the host environment, which is required for microbial growth and metabolism. Iron is an essential nutrient that is frequently depleted within the host due to sequestration by host proteins such as transferrin and lactoferrin. Siderophores, such as *Escherichia coli* enterobactin, have a high affinity for iron and can steal it from iron-binding proteins in the host. Iron acquisition is critical for pathogen survival and proliferation within the host (Meena et al., 2019).

8.2.7. Enzymes

In addition to aiding in invasion, other enzymes help with nutrient acquisition and immune evasion. Hemolysins are enzymes that lyse red blood cells, releasing hemoglobin, which can then be broken down to release iron (Chatterjee and Raval, 2019). Proteases degrade host proteins, providing nutrients while also deactivating immune components such as antibodies and complement proteins.

8.3. MICROBIAL TOXINS

Microbial toxins, primarily exotoxins and endotoxins, contribute to infection pathogenesis by causing direct tissue damage, disrupting normal cellular functions, and inducing inflammatory responses. Understanding the mechanisms of action and effects of these toxins is critical in microbiology, as they are frequently important predictors of disease severity and clinical outcomes.

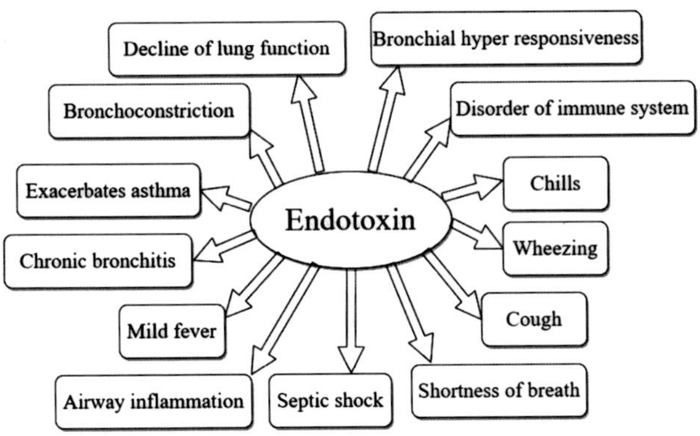

Figure 8.2. *The effects of endotoxins.*

Source: Chhetry BSK, Dewangan KN, Mahato DK, Kumar P. (2023). Endotoxins affecting human health during agricultural practices: An overview. Applied Chem., 3(1), 11–31. https://doi.org/10.3390/appliedchem3010002).

8.3.1. Exotoxins

Exotoxins are proteins secreted by various bacteria, with potent effects on host cells and tissues. These toxins are typically categorized based on their target or mechanism of action, such as cytotoxins, neurotoxins, and enterotoxins.

- **Cytotoxins:** Cytotoxins target and destroy host cells, causing tissue damage and aiding in the spread of infection. One example is *Staphylococcus aureus*-secerted alpha-toxin, a pore-forming cytotoxin that integrates into host cell membranes, forming transmembrane pores that cause cell lysis and death (Cheung et al., 2021). This action destroys tissues, contributing to the invasiveness of staphylococcal infections.
- **Neurotoxins:** Neurotoxins specifically target nerve cells, interfering with neurotransmission and causing neurological symptoms. The botulinum toxin produced by *Clostridium botulinum* is one of the most potent neurotoxins known. It blocks the release of acetylcholine at neuromuscular junctions, leading to flaccid paralysis, a condition characterized by muscle weakness and respiratory failure. Another example is the tetanus toxin from *Clostridium tetani*, which inhibits inhibitory neurotransmitters, resulting in spastic paralysis, where muscles are in a constant state of contraction (Fortier and Sekulovic, 2013).
- **Enterotoxins**: Enterotoxins attack the intestines, causing fluid secretion and diarrhea, which are typical symptoms of gastrointestinal infections. *Vibrio cholerae* produces cholera toxin, which is a classic example. It ADP-ribosylates the G protein involved in the regulation of adenylate cyclase in intestinal epithelial cells, leading to the overproduction of cyclic AMP (cAMP). This increase in cAMP levels causes excessive secretion of water and electrolytes into the intestinal lumen, resulting in severe diarrhea and dehydration. The enterotoxins produced by *Escherichia coli*, such as the heat-labile toxin (LT), function similarly, contributing to the pathogenesis of traveler's diarrhea (Murray et al., 2015).

8.3.2. Endotoxins

Endotoxins, as opposed to exotoxins, are essential elements of the outer membrane of Gram-negative bacteria. The most notable endotoxin is LPS, which is secreted during bacterial lysis or cell division. LPS is made up of three components: lipid A, the O-antigen, and the core oligosaccharide. It is lipid A that causes toxic effects (Madigan et al., 2018). When LPS enters the bloodstream, it triggers a robust immune response by binding to Toll-like receptor 4 (TLR4) on immune cells, leading to the activation of the nuclear factor kappa-light-chain-enhancer of activated B cells (NF-κB) pathway. This activation results in the production of pro-inflammatory cytokines, such as tumor necrosis factor-alpha (TNF-α) and interleukin-1 (IL-1), which can lead to systemic inflammation. In severe cases, this inflammatory response can cause septic shock, a life-threatening condition characterized by widespread tissue damage, organ failure, and death (Tay et al., 2016; Meena et al., 2019). The ability of endotoxins to induce such a profound inflammatory response underscores their importance in microbial pathogenicity.

8.3.3. Superantigens

Superantigens are a type of microbial toxin that bridges the gap between exotoxins and immunomodulators. These toxins, such as *Staphylococcus aureus*-caused toxic shock syndrome toxin-1 (TSST-1), bypass normal antigen processing and directly stimulate a large number of T cells by binding to major histocompatibility complex (MHC) class II molecules on antigen-presenting cells (APCs) and the variable region of the T-cell receptor (Fortier and Sekulovic, 2013). This causes a massive release of cytokines, known as a "cytokine storm," which can lead to toxic shock syndrome, which includes fever, rash, and multi-organ failure.

8.3.4. Hemolysins

Hemolysins are another type of exotoxin that targets red blood cells, causing them to lyse and release hemoglobin, which pathogens can use for iron. *Staphylococcus aureus*-produced alpha-hemolysin and streptolysin O of *Streptococcus pyogenes* are two examples of hemolysins that contribute to the virulence of these bacteria by damaging host cells and tissues, respectively (Cheung et al., 2021).

8.4. IMMUNE RESPONSES TO MICROBIAL INFECTIONS

The host immune response to infection is a highly coordinated and dynamic process that involves multiple layers of defense mechanisms aimed at recognizing, controlling, and eliminating invading pathogens. These immune responses are critical for maintaining homeostasis and preventing infections from spreading within the body. Innate and adaptive immunity are the primary types of immune responses that play distinct but complementary roles in combating infections (Tay et al., 2016).

8.4.1. Innate Immune Response

The innate immune response is the first line of defense against microbial infections, distinguished by its rapid onset and non-specific nature. This response includes physical barriers, cellular defenses, and soluble factors. The primary defense mechanisms are physical barriers such as the skin and mucous membranes, which prevent pathogens from entering the host. Furthermore, the secretion of antimicrobial peptides such as defensins and enzymes such as lysozyme directly targets and kills invading microorganisms (Thammavongsa et al., 2015).

Phagocytic cells such as macrophages, neutrophils, and dendritic cells play a role in innate immunity at the cellular level. These cells use PRRs, such as TLRs, to recognize PAMPs. Phagocytosis is the process by which phagocytes engulf and digest pathogens after they have been recognized. This action is critical for both immediate infection containment and the initiation of subsequent immune responses (Thammavongsa et al., 2015).

The complement system, which is a collection of plasma proteins, is another critical component of the innate immune response. The activation of the complement cascade results in pathogen opsonization, enhanced phagocytosis, and the formation of the membrane attack complex (MAC), which directly lyses pathogens (Tay et al., 2016). The innate immune response is thus critical for both immediate protection and shaping the adaptive immune response that follows.

8.4.2. Adaptive Immune Response

The adaptive immune response is highly specific and provides long-lasting immunity by forming immunological memory. Unlike innate immunity, the adaptive response takes longer to develop because it requires the activation and clonal expansion of lymphocytes in response to specific antigens. This response is mediated by B cells and T cells, both of which play distinct roles in infection defense (Fortier and Sekulovic, 2013).

B cells drive humoral immunity, which is the production of antibodies that specifically bind to antigens on pathogen surfaces. These antibodies neutralize pathogens, help them opsonize for phagocytosis, and stimulate the complement system. After antigen binding and T-cell stimulation, B cells differentiate into plasma cells that secrete large amounts of antibodies (Thammavongsa et al., 2015).

T cells are responsible for the majority of cell-mediated immunity. Helper T cells (CD4+) regulate the immune response by secreting cytokines that activate other immune cells such as B cells and cytotoxic T cells (CD8+). Cytotoxic T cells identify and eliminate infected cells by detecting antigens on MHC class

I molecules. This process is especially important in viral infections, where the elimination of infected host cells prevents the virus from spreading (Tay et al., 2016). The adaptive immune response not only eliminates the current infection, but it also establishes immunological memory, allowing for a faster and more effective response following exposure to the same pathogen.

8.4.3. Antigen Presentation

Antigen presentation is an important process that links innate and adaptive immunity, allowing the immune system to recognize and respond to pathogens effectively. APCs, which include dendritic cells, macrophages, and B cells, are critical in this process. APCs capture pathogens or their components, process them, and present peptide fragments on their surface that are bound to MHC molecules (Thammavongsa et al., 2015).

Antigen presentation involves two classes of MHC molecules: MHC class I and MHC class II. MHC class I molecules present endogenous antigens, which are typically derived from intracellular pathogens such as viruses to cytotoxic T cells (CD8+). This presentation activates cytotoxic T cells, which then target and eliminate infected cells (Meena et al., 2019). MHC class II molecules, on the other hand, present exogenous antigens to helper T cells (CD4+), which are typically from extracellular pathogens. This interaction is required for the activation of helper T cells, which then stimulate B cells to produce antibodies and macrophages to improve phagocytosis (Murray et al., 2015). Antigen presentation is thus critical to the activation and regulation of the adaptive immune response.

8.4.4. Inflammatory Response

An important feature of the immune system's reaction to an infection is the inflammatory response, which is typified by pain, swelling, heat, and redness at the infection site. Pro-inflammatory cytokines released by activated macrophages and other immune cells, including IL-1, IL-6, and TNF-α, start this response. According to Tay et al. (2016), these cytokines promote vascular permeability, which makes it easier for mediators, immune cells, and plasma proteins to reach the infection site.

Inflammation serves several functions in the immune response. It prevents the spread of infection by forming a physical barrier, attracts immune cells to the site of infection, and promotes pathogen and damaged tissue clearance. Excess or chronic inflammation, on the other hand, can cause tissue damage and is associated with a variety of inflammatory diseases. Thus, inflammation regulation is critical for balancing effective pathogen clearance and tissue preservation (Cheung et al., 2021).

8.4.5. Interferons and Antiviral Responses

IFNs are a type of cytokine that play an important role in the host's defense against viral infections. Infected cells produce Type I interferons (primarily IFN-α and IFN-β) in response to viral recognition. These interferons induce an antiviral state in neighboring cells by increasing the expression of interferon-stimulated genes (ISGs), which inhibit viral replication and promote viral RNA degradation (Madigan et al., 2018).

In addition to their direct antiviral effects, interferons enhance dendritic cells' antigen presentation capabilities, promoting the activation of cytotoxic T cells and the overall adaptive immune response. NK cells and T cells produce Type II interferon (IFN-γ), which activates macrophages and enhances their ability to phagocytose and destroy pathogens (Thammavongsa et al., 2015). Interferon production is therefore critical for controlling viral infections and orchestrating the overall immune response.

8.5. ANTIMICROBIAL RESISTANCE

Antimicrobial resistance (AMR) is a major challenge in modern medicine, reducing the efficacy of treatments for bacterial, fungal, viral, and parasitic infections. This phenomenon occurs when microorganisms develop mechanisms to survive exposure to antimicrobial agents that would normally inhibit or kill them (Madigan et al., 2018). AMR develops through a multifaceted process influenced by both genetic and environmental factors, posing a significant threat to global public health.

The following are key mechanisms leading to AMR:

8.5.1. Genetic Mutations and Horizontal Gene Transfer

The genetic basis of AMR frequently includes microbial DNA mutations that provide survival advantages under antimicrobial pressure. Spontaneous mutations can alter antibiotic target sites, reducing the drug's binding affinity and thus efficacy. Mutations in the gyrA gene in *Escherichia coli*, for example, can cause resistance to fluoroquinolones by affecting DNA gyrase, the enzyme that these antibiotics target (Morrison and Zembower, 2020).

In addition to mutations, horizontal gene transfer (HGT) plays an important role in the spread of resistance. HGT is the acquisition of resistance genes from other microorganisms via processes such as transformation, transduction, and conjugation. The transfer of plasmids carrying bla genes encoding β-lactamases has facilitated the widespread dissemination of resistance to β-lactam antibiotics, including penicillins and cephalosporins (Holmes et al., 2016). Enzymes hydrolyze β-lactam rings, making antibiotics ineffective.

8.5.2. Enzymatic Degradation of Antimicrobials

Certain microorganisms that are resistant to antibiotics generate enzymes that alter or break down these agents chemically. One well-known instance is how bacteria produce β-lactamases, which confer resistance to β-lactam antibiotics. The ability of extended-spectrum β-lactamases (ESBLs) to hydrolyze a variety of β-lactams, such as third-generation cephalosporins, presents a serious challenge in clinical settings (Morrison and Zembower, 2020).

Maier et al. (2009) have provided another example, which is the synthesis of enzymes known as aminoglycoside-modifying enzymes. These enzymes phosphorylate, acetylate, or adenylate aminoglycoside antibiotics, thereby impeding their ability to bind to their intended ribosomal sites and consequently decreasing their antimicrobial efficacy.

8.5.3. Alteration of Target Sites

Microorganisms can acquire resistance by changing the molecular targets of antimicrobial agents. This mechanism frequently involves mutations or post-translational modifications to target proteins. *Staphylococcus aureus* (MRSA) develops resistance to methicillin through the acquisition of the mecA gene, which encodes a penicillin-binding protein (PBP2a) with low affinity for β-lactam antibiotics. This modification enables bacteria to continue synthesizing cell walls in the presence of these drugs (Holmes et al., 2016).

8.5.4. Efflux Pumps and Reduced Permeability

Efflux pumps are transport proteins that actively expel antimicrobial agents from microbial cells, lowering their intracellular concentrations and thus their efficacy. These pumps are frequently encoded by resistance genes that can be acquired via HGT. For example, *E. coli's* AcrAB-TolC efflux pump contributes to resistance to a wide range of antibiotics, including tetracyclines and fluoroquinolones (Madigan et al., 2018).

Reduced permeability is another mechanism by which microbial cell wall or membrane changes prevent antimicrobial agents from entering. Gram-negative bacteria, for example, can change the outer membrane porins to reduce antibiotic uptake, contributing to resistance (Holmes et al., 2016; Morrison and Zembower, 2020).

8.6. EMERGING INFECTIOUS DISEASES

Emerging infectious diseases (EIDs) are those that have recently developed in a community or have existed but are quickly expanding in incidence or geographic range. These diseases represent substantial challenges to global health because of their potential for widespread effect and the complexity associated with their management. EIDs have historically had a significant impact on human society, as demonstrated by pandemics such as the 1918 influenza outbreak and more contemporary events such as the COVID-19 pandemic. The research of EIDs in microbiology is critical for understanding their origins, causes, and effects, which allows for the creation of effective management and control techniques.

8.6.1. Drivers of the Emergence of Infectious Diseases

The genesis of infectious diseases is influenced by a complex interaction of ecological, environmental, and socioeconomic factors. Environmental changes, such as climate change and deforestation, modify pathogen and vector habitats, increasing the risk of disease transmission. Rising temperatures, for example, might cause vector-borne diseases such as malaria and dengue fever to spread to new locations as mosquitos and ticks migrate (McCloskey et al., 2014).

The emergence of diseases is also significantly influenced by human activity. As seen by the worldwide spread of SARS-CoV-2, globalization, and increased human mobility allow infectious pathogens to spread quickly across national boundaries. Since intimate human-animal interaction raises the risk of cross-species transmission, urbanization, and intensive agriculture are factors in the rise of zoonotic illnesses. Furthermore, ecosystems are disrupted by changes in land use and habitat fragmentation, which increases interactions between livestock, humans, and animals (Lorusso, 2020). Genetic alterations and microbial evolution are the underlying causes of illness onset. Genetic recombination and mutations occur in pathogens, and these changes might improve the pathogen's virulence, resistance to therapy, or capacity to infect new hosts. For instance, the widespread use of antibiotics has created selective pressure that has led to the emergence of antibiotic-resistant bacterial strains, such as Methicillin-resistant *Staphylococcus aureus* (MRSA) (McCloskey et al., 2014).

8.6.2. Mechanism of Disease Transmission

Controlling EIDs requires an understanding of the illness transmission processes. There are three types of transmission pathways: vector-borne, indirect, and direct. Direct transmission occurs when an infected person comes into direct physical touch with a vulnerable host. Examples include respiratory infections that spread by droplets, such as tuberculosis and influenza.

Fomites, or intermediary items or surfaces, are the conduits through which indirect transmission happens. Pathogens have the ability to cling to surfaces and spread when a vulnerable person comes into contact with these contaminated items. This method of transmission matters for illnesses like the norovirus and several viruses linked to healthcare (Boyce and Pittet, 2002).

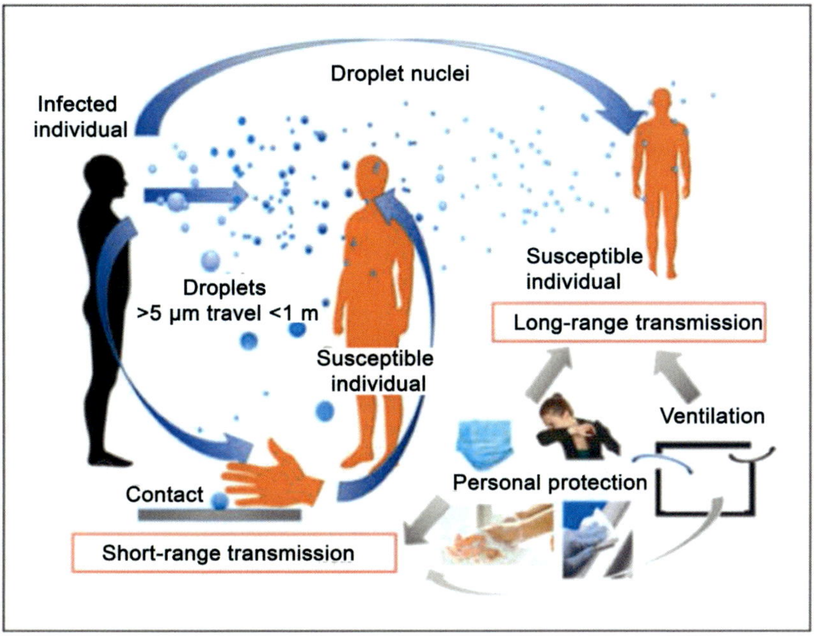

Figure 8.3. *Mode of transmission of SARS-CoV-2.*

Source: Delikhoon M, Guzman MI, Nabizadeh R, Norouzian Baghani A. (2021). Modes of transmission of severe acute respiratory syndrome-coronavirus-2 (SARS-CoV-2) and factors influencing on the airborne transmission: A review. International Journal of Environmental Research and Public Health, 18(2):395. https://doi.org/10.3390/ijerph18020395.

Vector-borne transmission occurs when organisms such as mosquitoes, ticks, and fleas transport infections between hosts. Vectors play an important part in the lifecycle of many infectious pathogens, including *Plasmodium* spp., which cause malaria and are transmitted by Anopheles mosquitos (Lorusso, 2020). Understanding vector ecology and control is critical for treating diseases, such as dengue fever, the Zika virus, and Lyme disease.

8.6.3. Surveillance and Detection

Effective surveillance and detection are critical for prompt identification and response to EIDs. Surveillance systems monitor illness incidence and outbreaks using a variety of approaches, including routine reporting from healthcare facilities, laboratory testing, and syndromic surveillance, which tracks health indicators such as emergency room visits (Morse, 2012).

Diagnostic techniques have evolved dramatically, improving our ability to detect and diagnose diseases. Molecular approaches, such as PCR and NGS, are highly sensitive and specific for detecting pathogens and their genetic material. Serological tests, which identify antibodies or antigens in the blood, can also help diagnose viral infections and understand immunological responses.

The integration of surveillance data with GIS enables spatial study of illness patterns, hence facilitating targeted interventions. Early warning systems, which employ data from a variety of sources, including environmental monitoring and predictive modeling, can detect and prevent outbreaks before they become widespread (Morse, 2012).

8.6.4. Prevention and Control

Preventing and controlling EIDs requires a multifaceted approach that includes vaccination, public health interventions, and education. Vaccination is one of the most effective ways to avoid infectious diseases. Vaccines can prevent the spread of diseases such as measles, influenza, and hepatitis B, and therefore are essential for controlling outbreaks (Lorusso, 2020).

Quarantine, travel limitations, and the promotion of basic cleanliness are all examples of public health actions. During epidemics, quarantine and isolation can assist limit the transmission of infectious diseases, whereas travel restrictions help prevent international spread. Promoting hand cleanliness, healthy dietary practices, and vector control methods are critical to minimizing transmission in communities.

Education and awareness campaigns are critical in teaching the public about illness-preventive techniques and promoting adherence to health recommendations. Collaboration among governments, international organizations, and communities is critical for implementing effective preventative and control strategies (Lorusso, 2020).

8.7. NOSOCOMIAL INFECTIONS

Nosocomial infections, commonly known as hospital-acquired infections (HAIs), are diseases that patients contract while being treated for other conditions in a healthcare setting. These infections are a leading cause of morbidity, mortality, and increased healthcare expenses worldwide.

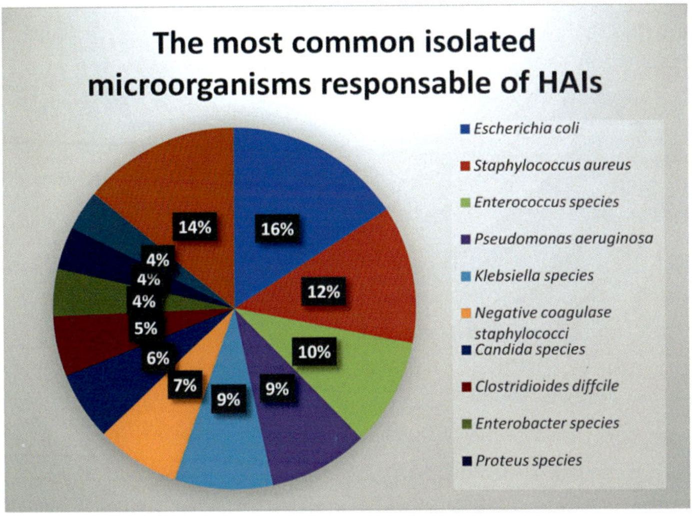

Figure 8.4. *Representation of the most common isolated bacteria responsible for HAI in EU.*

Source: Szabó S, Feier B, Capatina D, Tertis M, Cristea C, Popa A. (2022). An overview of healthcare associated infections and their detection methods caused by pathogen bacteria in Romania and Europe. Journal of Clinical Medicine, 11(11):3204. https://doi.org/10.3390/jcm11113204.

Despite advances in medical procedures and infection control, nosocomial infections continue to pose a significant concern for healthcare systems. Their prevalence highlights the importance of continued surveillance, prevention, and control efforts to protect patients, especially those who are most vulnerable.

8.7.1. Etiology of Nosocomial Infections

Numerous pathogens, including bacteria, viruses, fungi, and parasites, can cause nosocomial infections. The most prevalent bacterial pathogens are *Escherichia coli*, *Pseudomonas aeruginosa*, *Klebsiella pneumoniae*, and *Staphylococcus aureus*, especially MRSA. These microorganisms are frequently linked to illnesses such as bloodstream, urinary tract, pneumonia, and surgical site infections. *Staphylococcus aureus* is particularly well-known for causing surgical site infections since it is often acquired through the skin or contaminated surgical instruments (Sikora and Zahra, 2024).

Escherichia coli is a common cause of urinary tract infections and can be acquired by catheterization. *Pseudomonas aeruginosa*, notorious for its drug resistance, is frequently found in respiratory and wound infections, particularly in immunocompromised people. Fungal pathogens, such as Candida species, can potentially be dangerous, especially in patients with weaker immune systems. These pathogens can be transmitted by contaminated medical devices, healthcare personnel' hands, environmental surfaces, and other patients (Sikora and Zahra, 2024).

8.7.2. Risk Factors and Vulnerable Population

The chance of contracting nosocomial infections is increased by multiple factors. These include extended hospital stays, invasive medical treatments including mechanical breathing or catheterization, surgical procedures, and the use of broad-spectrum antibiotics. Since intrusive devices are used often and patients' illnesses are severe, patients in intensive care units (ICUs) are especially vulnerable. Treatment for healthcare-associated infections is further complicated by the overuse and misuse of antibiotics, which also leads to the rise of microbes resistant to several drugs (Sikora and Zahra, 2024).

The elderly, infants, immunocompromised people (such as those receiving chemotherapy or organ transplants), and patients with long-term illnesses like diabetes or chronic obstructive pulmonary disease (COPD) are among the vulnerable groups. These individuals have less developed immune systems, which makes it harder for their bodies to fight off illnesses. Because surgical patients may come into contact with infections during the process, their risk is also elevated (Meena et al., 2019).

8.7.3. Surveillance and Detection

Effective surveillance is essential for early detection and treatment of nosocomial infections. Surveillance is the systematic collection, analysis, and interpretation of health data to track the prevalence of HAIs in healthcare settings. This technique is critical for detecting infection trends, and outbreaks, and assessing the efficacy of infection control interventions.

Modern surveillance systems combine manual reporting with computerized electronic health records (EHRs). The combination of microbiological data and clinical information enables real-time detection of infection clusters and trends. The Centers for Disease Control and Prevention (CDC) offers recommendations and tools for HAI surveillance, including the National Healthcare Safety Network (NHSN), which is widely utilized in the US. Active surveillance cultures, particularly in high-risk environments such as ICUs, aid in identifying carriers of multidrug-resistant organisms (MDROs), allowing for focused infection management measures (Lorusso, 2020).

8.7.4. Prevention and Control

A multimodal strategy is needed for the prevention and treatment of nosocomial infections. This includes wearing personal protective equipment (PPE), sterilizing medical equipment, adhering strictly to infection control regulations, and practicing good hand hygiene. In order to reduce the risk of illness transmission, healthcare facilities need to take both standard and transmission-based precautions.

The single most crucial step in stopping the transmission of illnesses is practicing good hand hygiene. For routine hand antisepsis, the World Health Organization (WHO) advises using alcohol-based hand rubs since they are quicker and more efficient than soap and water. Reducing surface and environmental contamination also requires appropriate waste disposal, environmental cleaning, and disinfection.

Antibiotic stewardship programs play a vital role in preventing the development of antibiotic-resistant strains by ensuring the appropriate use of antibiotics. Education and training of healthcare personnel on infection control practices are essential to maintaining a culture of safety within healthcare facilities (Madigan et al., 2018).

8.7.5. Outbreak Management and Response

In order to stop the spread of nosocomial infection outbreaks, quick action and coordination are needed. Enhanced surveillance and laboratory testing are used to identify and confirm the outbreak, which is the first step in outbreak management. Infection control teams are required to carry out a comprehensive investigation to determine the source and mechanism of transmission upon confirmation of an outbreak.

Isolating afflicted individuals, improving hand hygiene, cleaning the surroundings, and examining the usage of medical instruments are a few examples of control strategies. To contain the outbreak, it could occasionally be required to segregate patients and personnel or temporarily close the impacted wards. In order to control the situation and preserve public confidence in the healthcare institution, communication with patients, healthcare workers, and the general public is essential. Documentation and analysis of the outbreak response provide valuable lessons for preventing future occurrences. Continuous evaluation and improvement of infection control practices are essential in safeguarding patients from nosocomial infections (Lorusso, 2020).

8.8. ZOONOTIC INFECTIONS

Zoonotic illnesses, often known as zoonoses, are diseases that naturally spread from animals to people. These diseases are caused by a variety of pathogenic microorganisms, including viruses, bacteria, parasites, and fungi, which can live in animal hosts and then infect humans. Zoonotic infections have a substantial impact on microbiology since they account for about 60% of all infectious diseases that afflict people. The complex connections between humans, animals, and the environment accelerate the spread of these infections, making zoonoses an important topic of study in epidemiology and public health.

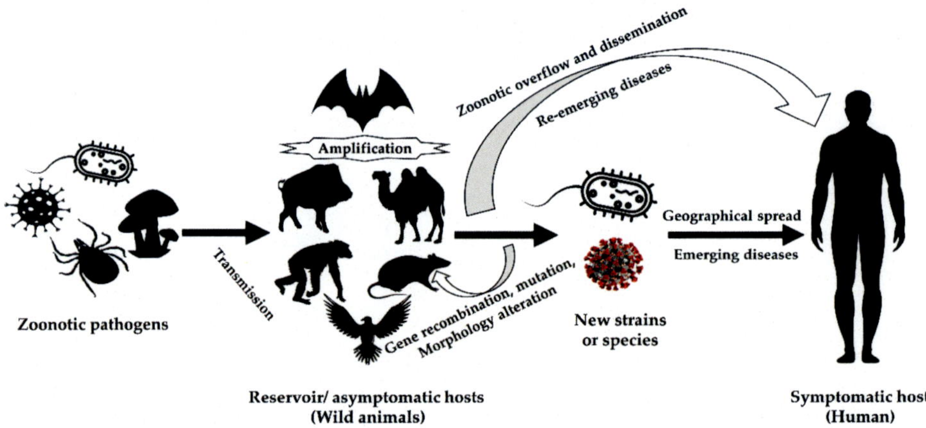

Figure 8.5. *Emerging zoonotic diseases.*

Source: Rahman MT, Sobur MA, Islam MS, Ievy S, Hossain MJ, El Zowalaty ME, Rahman AT, Ashour HM. (2020). Zoonotic diseases: Etiology, impact, and control. Microorganisms, 8(9), 1405. https://doi.org/10.3390/microorganisms8091405.

8.8.1. Pathogens and Reservoir Host

Zoonotic diseases are caused by a variety of pathogens, including viruses, bacteria, fungi, and parasites. Notable zoonotic pathogens include the rabies virus, the Ebola virus, and the new coronavirus (SARS-CoV-2). Bacterial zoonoses include diseases like leptospirosis produced by Leptospira species and anthrax, which is caused by *Bacillus anthracis*. Toxoplasmosis, caused by the protozoan *Toxoplasma gondii*, and echinococcosis, caused by *Echinococcus* tapeworms, are examples of parasitic zoonotic diseases. Fungal zoonoses, which are less prevalent, include illnesses such as ringworm caused by *Microsporum* and *Trichophyton* species (Lorusso, 2020).

Reservoir hosts are animals that carry these infections but do not exhibit illness symptoms. Dogs, cats, and livestock are common reservoir hosts, as are bats, rats, and birds. For example, bats are known reservoirs for viruses like Ebola and coronaviruses, whereas rodents are reservoirs for hantaviruses and the bacteria that cause plague (*Yersinia pestis*). The transmission of these viruses from reservoir hosts to humans can result in large epidemics, as demonstrated by the global spread of COVID-19 (Maier et al., 2009).

8.8.2. Mechanism of Disease Transmission

The processes of zoonotic disease transmission are numerous and can be divided into various categories:

- **Direct Transmission**: Refers to direct physical contact between an infected animal and a human. Examples include bites, scrapes, and direct animal handling. Rabies, for example, is transmitted by bites or scratches with infected animals' saliva (Warrell, 2018).
- **Indirect Transmission**: Pathogens can spread indirectly through contaminated settings or things. This involves handling or ingesting contaminated food or water. Salmonella, for example, can spread from infected meat or eggs, resulting in foodborne infections (Khan et al., 2018).
- **Vector-Borne Transmission**: Ticks, mosquitoes, and fleas help infections spread from animals to humans. Ticks transmit Borrelia burgdorferi, the causal agent of Lyme disease, whereas Anopheles mosquitoes transmit Plasmodium spp., which causes malaria (Madigan et al., 2018; Meena et al., 2019).

8.8.3. Risk Factors and Vulnerable Populations

A range of risk factors make some populations more vulnerable to zoonotic illnesses. Those who work closely with animals, such as farmers, veterinarians, and wildlife handlers, are more vulnerable. Furthermore, people who reside in places with a lot of diversity in the fauna or in locations with inadequate cleanliness and sanitation standards are particularly vulnerable. People with impaired immune systems, such as the elderly, small children, and those suffering from long-term illnesses, are also more vulnerable to zoonotic diseases.

Poverty and a lack of access to healthcare increase the risk of zoonotic illnesses. Many underdeveloped countries lack the necessary infrastructure and public health resources to effectively prevent and control chronic diseases. Climate change and environmental degradation also influence the emergence and re-emergence of zoonotic diseases by changing the habitats of reservoir hosts and vectors(Morse, 2012).

8.8.4. Surveillance and Diagnosis

Effective surveillance and diagnostics are critical for early detection and management of zoonotic infections. Surveillance systems monitor the prevalence and spread of zoonotic illnesses in both human and animal populations. These systems rely on coordination among public health officials, veterinary services, and environmental organizations. Advanced diagnostic techniques, like PCR, serological testing, and next-generation sequencing (NGS), allow for reliable detection of zoonotic pathogens.

Early detection is critical for effective treatment and containment of zoonotic outbreaks. Integrated approaches, such as the One Health paradigm, stress the interdependence of human, animal, and environmental health, encouraging comprehensive surveillance strategies. However, issues persist in resource-constrained environments, where diagnostic facilities and surveillance networks may be insufficient (Morse, 2012).

8.8.5. Prevention and Control Measures

Preventing and controlling zoonotic infections necessitates a multifaceted approach that combines public health interventions, veterinary measures, and environmental management. Vaccination efforts, public awareness campaigns, and the promotion of good hygiene habits are examples of public health measures for reducing transmission risk. For example, vaccinating dogs against rabies is an important step in preventing the virus from spreading to humans.

Veterinary measures involve monitoring and regulating animal health, particularly in cattle and domestic animals. Quarantine protocols, routine health checks, and the use of antibiotics or antiparasitic treatments all serve to manage zoonoses in animal populations. Environmental management, such as vector control programs, adequate waste disposal, and safe food handling procedures, all help to reduce the spread of zoonotic diseases (Madigan et al., 2018; Meena et al., 2019).

8.9. VACCINES AND ANTIMICROBIAL THERAPIES

Vaccines and antimicrobial medicines are essential interventions in modern medicine, preventing and treating infectious illnesses. Vaccines activate the immune system to guard against certain pathogens, whereas antimicrobial medicines target and kill infectious organisms such as bacteria, viruses, fungi, and parasites. Both approaches have had a significant impact on public health, lowering the incidence and fatality rates of infectious diseases. Continued progress in these disciplines is critical for tackling new diseases and resistance.

8.9.1. Types of Vaccines

Vaccines can be generally classified into numerous categories, each using a distinct mechanism to induce an immune response.

- **Inactivated Vaccinations**: This includes germs that have been killed or inactivated, making them incapable of causing disease. Examples include polio and hepatitis A vaccinations.
- **Live Attenuated Vaccinations**: These vaccinations use live pathogens that have been weakened so that they cannot infect healthy people. They often give long-term protection, as observed with the MMR vaccine.
- **Subunit, Recombinant, and Conjugate Vaccines**: These are vaccines that use specific pathogen components-such as proteins, carbohydrates, or capsids—to elicit an immune response. Examples include the HPV vaccine and the hepatitis B vaccine.
- **Toxoid Vaccines**: These vaccines are manufactured from inactivated toxins produced by the disease, allowing the immune system to attack the toxin's detrimental effects rather than the pathogen itself. Toxoid vaccinations include diphtheria and tetanus.
- **mRNA Vaccines**: A recent technique, mRNA vaccines, such as the COVID-19 vaccines from Pfizer-BioNTech and Moderna, employ messenger RNA to guide cells to make a protein that causes an immunological reaction (Lorusso, 2020).
- **Viral Vector Vaccines**: These vaccines use a modified virus (rather than the pathogen itself) to transport genetic information into cells, inducing an immune response. Examples include the Ebola vaccine and Johnson and Johnson's COVID-19 vaccine (World Health Organization, 2014).

Each form of vaccination has advantages and disadvantages, and the type chosen is determined by criteria such as the pathogen's nature, the target population, and available technology.

8.9.2. Mechanism of Antimicrobial Therapies

Antimicrobial medicines work by targeting microorganisms' vital activities, preventing their development or death. The mechanisms differ based on the type of antibacterial agent.

- **Antibiotics** inhibit cell wall synthesis (e.g., penicillins), disrupt protein synthesis (e.g., tetracyclines), interfere with nucleic acid replication and transcription (e.g., fluoroquinolones), or compromise cell membrane integrity (e.g., polymyxins) (Hewitt, 2004).

- **Antivirals**: These drugs target certain stages of viral propagation. For example, nucleoside analogs such as acyclovir suppress herpesvirus DNA replication, but protease inhibitors used in HIV therapy hinder viral protein maturation.
- **Antifungals** disturb the cell membrane (azoles decrease ergosterol synthesis) or hinder cell division (griseofulvin).
- **Anti-parasitics**: These drugs target parasitic infections and may work by interfering with energy production, nutrient uptake, or other critical functions of the parasite (Hewitt, 2004).

8.9.3. Resistance to Antimicrobial Therapies

AMR arises when germs develop methods to withstand the effects of medications that previously killed or reduced their growth. This resistance can occur through a variety of mechanisms:

- **Genetic Mutations**: Changes in microbial DNA can alter medication targets, rendering antibiotics useless.
- **Horizontal Gene Transfer**: Bacteria can acquire resistance genes from other bacteria by conjugation, transformation, or transduction (Boto, 2010).
- **Efflux Pumps**: Some bacteria produce efflux pumps that aggressively remove antimicrobial substances from the cell.
- **Enzymatic Degradation**: Certain bacteria develop enzymes, such as beta-lactamases, which breakdown antibiotics before they can take effect.

AMR is a developing global issue that complicates infection treatment while also increasing morbidity, death, and healthcare expenditures (Morrison and Zembower, 2020).

8.9.4. Vaccination Strategies and Public Health Implications

- **Vaccine Schedules and Extent of Coverage:** The goals of vaccination schedules are to maximize the benefits to public health while ensuring timely protection. Immunizations against measles, polio, and influenza are part of routine immunization schedules that are advised for adults, children, and babies. For herd immunity to develop, which protects the unvaccinated and slows the spread of disease, high vaccination coverage is essential.
- **Vaccine Safety and Efficacy:** Whereas vaccine effectiveness gauges a vaccine's performance in real-world situations, vaccine efficacy describes how well a vaccination prevents disease in optimal

circumstances. Vaccines are monitored for safety by post-marketing surveillance to make sure unanticipated side effects don't occur. Sustaining public confidence requires thorough safety assessments and ongoing oversight.

- **Impact on the Rate of Illness:** Smallpox has been eradicated thanks in large part to vaccination, which has also caused a notable drop in many other infectious diseases. To combat new threats like COVID-19 and control and eventually eradicate other illnesses like polio and measles, vaccination campaigns must continue (World Health Organization, 2014).

8.9.5. Future Directions in Vaccines and Antimicrobial Therapies

- **Technological Advances in Vaccines**: Vaccines targeting several strains or even distinct pathogens with broad protection are being developed as universal and nanoparticle vaccines. The development of adjuvants and better delivery methods could improve vaccination durability and responses (Gordon, 2020).
- **Novel Antimicrobial Agents**: Baciophage therapy and antimicrobial peptides are two potential alternatives to conventional medications being investigated in the field of novel antimicrobial agents research. These methods provide prospective remedies for treating infections with few available therapeutic alternatives and combating resistant microorganisms.
- **Opportunities and Challenges**: The persistent problem of antibiotic resistance and the introduction of novel diseases necessitate constant innovation and international cooperation. Investing in research, bolstering public health infrastructures, and encouraging global collaboration are all necessary to address these issues (World Health Organization, 2014).

8.10. PUBLIC HEALTH IMPLICATIONS

The public health implications of microbiology examine how microbial diseases affect population health and the techniques utilized to ameliorate these effects. This discipline investigates disease burden, new infections, antibiotic resistance, vaccination, and infection prevention strategies. Understanding these implications is critical for creating effective public health policies, improving disease prevention, and assuring equal access to healthcare interventions.

8.10.1. Disease Burden and Epidemiology

The disease burden refers to the overall impact of health problems on a population, including morbidity and mortality. Epidemiology gives insights into illness distribution and determinants, which help to guide public health initiatives. Infectious diseases like influenza, TB, and HIV/AIDS continue to have a significant impact on world health. According to the WHO, infectious diseases account for over 30% of the global disease burden, with emerging pathogens and resistant strains worsening the situation (World Health Organization, 2014). Epidemiological statistics are critical for identifying high-risk populations, comprehending illness patterns, and implementing effective prevention strategies.

8.10.2. Impact of Emerging Infectious Diseases

The term "EIDs" refers to newly or re-emerging illnesses with a growing geographic range or incidence. Recent outbreaks like COVID-19 and Ebola provide evidence of the significant effects that EIDs can have on public health. With over 600 million confirmed cases and severe economic and social disruptions, the COVID-19 pandemic, which was brought on by SARS-CoV-2, has brought attention to the world's susceptibility to novel infections. The formation of EIDs can be attributed to various factors such as environmental changes, human-animal interactions, and globalization. Changes in climate and urbanization modify the habitats of pathogens and raise the possibility of zoonotic spillovers. To manage EIDs and lessen their negative effects on public health, surveillance, and early detection are essential (Morse, 2012).

8.10.3. Antimicrobial Resistance and Its Consequences

AMR arises when germs adapt to withstand the effects of previously effective antibiotics. AMR is a major public health problem because it affects the efficacy of antibiotics, antivirals, antifungals, and anti-parasitics. This resistance can result in longer hospital stays, higher medical costs, and higher mortality. Resistance mechanisms include the creation of drug-degrading enzymes, drug target alteration, and enhanced drug efflux (World Health Organization, 2014). The emergence of multidrug-resistant bacteria, such as MRSA and Carbapenem-resistant Enterobacteriaceae (CRE), challenges treatment and control efforts (World Health Organization, 2014). Addressing AMR necessitates a holistic approach that includes stewardship programs, the discovery of novel antimicrobials, and international collaboration to deploy effective surveillance and response measures.

8.10.4. Vaccinations and Immunization Programs

Vaccinations are a critical component of public health, considerably lowering the incidence of infectious diseases. Vaccines stimulate the immune system to recognize and combat infections without producing disease. Successful immunization programs have resulted in the eradication of smallpox and major decreases in illnesses like polio and measles. The Centers for Disease Control and Prevention (CDC) highlights the significance of maintaining high vaccination rates to avoid outbreaks and build herd immunity. Vaccination challenges include vaccine reluctance, supply chain difficulties, and guaranteeing access among underprivileged groups. Addressing these difficulties through public education, equitable distribution, and strong vaccination programs is critical to maintaining public health (World Health Organization, 2014).

8.10.5. Infection Control and Prevention Measures

To stop the spread of infectious diseases and safeguard the public's health, infection control and prevention are crucial. Hygiene habits, which are essential for halting the spread of disease, include handwashing and adequate sanitation. Strict infection control procedures are essential in healthcare settings. These procedures include sterilizing medical equipment, using PPE, and adhering to infection management standards. To reduce risks and guarantee safety, the WHO offers guidelines for infection prevention and control (IPC) in a variety of contexts (Lorusso, 2020). Effective infection control programs also include immunization, surveillance, and outbreak management to quickly address and contain new threats.

8.10.6. Global Health and Equity Issues

Issues related to global health and equity bring to light differences in health outcomes and access to medical care across various communities. Communities that are marginalized and have low incomes frequently have greater rates of disease and have less access to resources for treatment and prevention. The international health community understands that resolving these disparities is essential to enhancing general health outcomes. By encouraging universal health coverage, enhancing access to immunizations and treatments, and fortifying health systems in underprivileged areas, initiatives like the Global Fund and the WHO's Health Equity Framework seek to alleviate gaps (World Health Organization, 2014). Coordinated efforts are needed to address socioeconomic determinants of health, guarantee equitable access to healthcare, and assist vulnerable groups in controlling and preventing infectious diseases to achieve health equity.

CHAPTER 9

MICROBIAL BIOREMEDIATION

Contents

Microbial bioremediation is an advanced ecological process that uses microorganisms' natural metabolic capabilities to degrade, detoxify, or transform pollutants in the environment. This process has received a lot of attention as a sustainable and cost-effective way to deal with environmental contamination, especially in soil, water, and sediment matrices. Microbial bioremediation is based on bacteria, fungi, and other microorganisms' ability to metabolize various organic and inorganic compounds, often converting them into less harmful substances or completely mineralizing them into carbon dioxide, water, and biomass (Malik and Goyal, 2022).

The selection of appropriate microbial strains, the optimization of environmental conditions to increase microbial activity, and an understanding of contaminant bioavailability are all important concepts in microbial bioremediation. Furthermore, the process may include strategies such as bioaugmentation, which introduces specific microbial strains to accelerate degradation, and biostimulation, which involves the addition of nutrients or other growth-promoting factors to increase the activity of the native microbial population (Tan et al., 2023).

Scholarly research has demonstrated the potential of microbial bioremediation in addressing a wide range of pollutants, including hydrocarbons, heavy metals, and emerging contaminants, emphasizing its importance in advancing environmental restoration efforts. This method is becoming more widely recognized as an essential component of long-term environmental management (Maier et al., 2009; Malik and Goyal, 2022).

9.1. BIODEGRADATION PROCESS

The biodegradation process is a series of metabolic steps in which microorganisms convert environmental pollutants into less harmful or inert compounds. This process is critical in the natural attenuation of contaminants, especially in soil and water environments, and is influenced by a variety of factors such as microbial community dynamics, environmental conditions, and the nature of the contaminants themselves (Malik and Goyal, 2022).

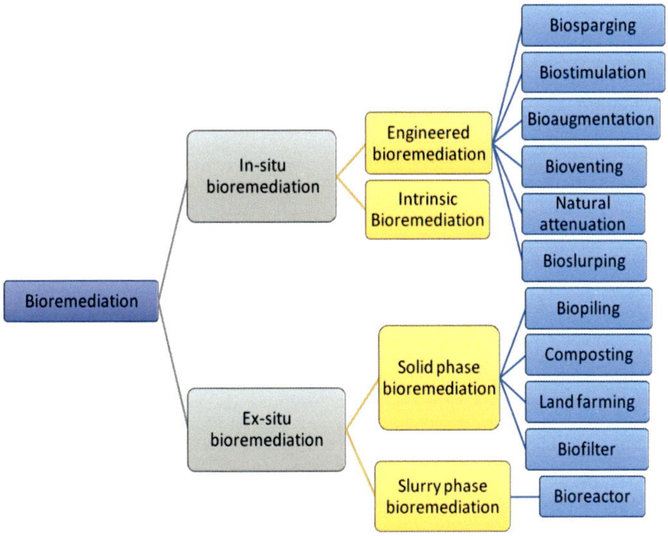

Figure 9.1. *Schematic representation of the different steps involved in biodegradation.*

Source: Bala S, Garg D, Thirumalesh BV, Sharma M, Sridhar K, Inbaraj BS, Tripathi M. (2022). Recent strategies for bioremediation of emerging pollutants: A review for a green and sustainable environment. Toxics, 10(8):484. https://doi.org/10.3390/toxics10080484.

9.1.1. Bioavailability

Bioavailability is a critical step in the microbial biodegradation process during bioremediation because it determines how accessible contaminants are to microbial cells for degradation. The physicochemical properties of the contaminant, such as solubility, hydrophobicity, and molecular structure, as well as the environmental matrix in which it resides, all influence bioavailability. Hydrophobic organic contaminants, such as polycyclic aromatic hydrocarbons (PAHs), have a low bioavailability because they bind strongly to soil particles or sediment, limiting their accessibility to microorganisms (Rajendran and Gunasekaran, 2019). Contaminants' bioavailability can be increased by using surfactants, biostimulation, or physical methods such as soil washing to mobilize pollutants and increase their interaction with microbial cells.

9.1.2. Uptake/Transport

Once contaminants are bioavailable, their absorption or transport into microbial cells is critical for biodegradation. The transportation mechanisms can be passive or active. The concentration gradient of the contaminant across the cell membrane drives passive transport via diffusion. This is common for small, hydrophobic molecules that can easily pass through the lipid bilayer of microbial cell membranes. Active transport, on the other hand, necessitates the use of energy and specific transport proteins that allow contaminants to move against a concentration gradient. Active transport is often required for larger, complex molecules that are not easily diffusible (Ławniczak et al., 2020). The physiological state of the cell and the availability of nutrients both influence uptake efficiency, which can either enhance or inhibit the transport process.

9.1.3. Enzymatic Transformation

Following the uptake of contaminants, microbial cells begin enzymatic transformation, a critical step that involves the metabolic breakdown of pollutants via a series of enzymatic reactions. Specific enzymes produced by microorganisms, tailored to the structure of the contaminant, catalyze these reactions. There are two main pathways for transformation: oxidative metabolism and reductive metabolism. In oxidative metabolism, oxygen serves as an electron acceptor, facilitating the degradation of organic contaminants via aerobic respiration, which is common for hydrocarbons and pesticides. Reductive metabolism occurs in anaerobic conditions and uses alternative electron acceptors such as nitrate, sulfate, or carbon dioxide, making it suitable for contaminants like chlorinated solvents (Ławniczak et al., 2020). These enzymatic transformations convert contaminants into intermediate metabolites, which are further processed by the cell.

9.1.4. Primary Degradation and Mineralization

Primary degradation is the initial breakdown of a contaminant into simpler, less complex intermediate compounds. This step is critical for reducing the toxicity of the pollutant. Mineralization occurs after primary degradation, during which these intermediates are further decomposed into inorganic substances like carbon dioxide, water, and mineral salts. Complete mineralization is the desired outcome in bioremediation because it completely removes the contaminant from the environment (Malik and Goyal, 2022). However, in some cases, partial degradation can result in the formation of more persistent or toxic byproducts, necessitating close monitoring of the biodegradation process.

9.1.5. Co-Metabolism

Co-metabolism is a phenomenon in which the presence of another substrate that serves as a secondary source of energy or carbon for microorganisms enhances the degradation of a primary contaminant. This process is especially important when dealing with recalcitrant or xenobiotic compounds, which may be difficult to metabolize on their own. During co-metabolism, enzymes produced to degrade the secondary substrate also transform the primary contaminant, often leading to its eventual breakdown. Co-metabolism degradation products, on the other hand, are not always fully mineralized, and additional steps may be required to ensure complete remediation (Rajendran and Gunasekaran, 2019).

9.1.6. Environmental Factors and Optimization

Environmental factors such as pH, temperature, oxygen availability, and nutrient concentration have a significant impact on the rate and efficiency of microbial biodegradation. Optimal conditions for microbial growth and enzyme activity are required to maximize the biodegradation process. For example, most biodegradation processes work best at neutral pH and moderate temperatures, though some extremophiles can function in more extreme conditions (Maier et al., 2009). Aerobic degradation requires oxygen, whereas anaerobic degradation requires alternative electron acceptors. Microbial metabolism and enzyme production require nutrients like nitrogen, phosphorus, and trace elements. To optimize these conditions, strategies like biostimulation, which involves the addition of nutrients or electron acceptors, and bioaugmentation, where specific microbial strains are introduced, can be employed to enhance the biodegradation process (Rajendran and Gunasekaran, 2019).

9.2. BIOREMEDIATION STRATEGIES

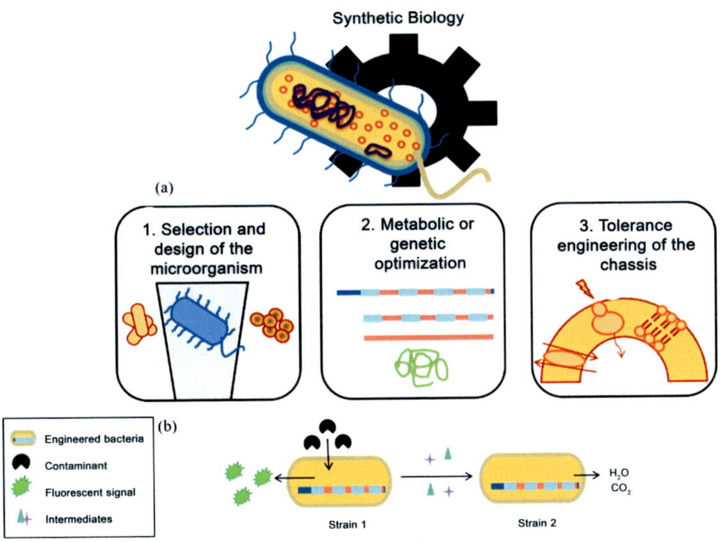

Figure 9.2. *The strategies of synthetic biology applicable for bioremediation.*

Source: Jiménez-Díaz V, Pedroza-Rodríguez AM, Ramos-Monroy O, Castillo-Carvajal LC. (2022). Synthetic biology: A new era in hydrocarbon bioremediation. Processes, 10(4), 712. https://doi.org/10.3390/pr10040712.

The most important bioremediation strategies, which use microorganisms' metabolic capabilities to reduce environmental contamination, are as follows:

9.2.1. Natural Attenuation

Natural attenuation is the process by which indigenous microorganisms degrade, disperse, or stabilize environmental contaminants without the need for human intervention. This passive strategy relies on microbial communities' inherent ability to metabolize pollutants under current environmental conditions. Natural attenuation is frequently considered in situations where contaminant levels are low and the environmental impact is minimal. It is usually monitored over time to ensure that the contaminants are effectively reduced to non-toxic levels (Verma and Kuila, 2019). While cost-effective, this approach necessitates a thorough understanding of the site's conditions and contaminant' behavior, as it may be slower and less predictable than active remediation strategies.

9.2.2. Bio-Stimulation

Bio-stimulation is the process of increasing microbial activity in contaminated environments by adding nutrients, electron donors, or other growth-promoting factors. This strategy aims to improve environmental conditions so that the existing microbial population can degrade contaminants more efficiently. Bio-stimulation is especially useful in nutrient-limited environments, such as petroleum-contaminated soils, where the addition of nitrogen, phosphorus, and oxygen can significantly speed up biodegradation (Rajendran and Gunasekaran, 2019). Bio-stimulation's success is dependent on carefully balancing nutrient inputs and monitoring microbial activity to avoid unintended consequences, such as the proliferation of non-degrading microbial species.

9.2.3. Bioaugmentation

Bioaugmentation is the introduction of specific microbial strains or consortia into a contaminated site in order to improve pollutant biodegradation. This method is commonly used when the indigenous microbial community lacks the metabolic pathways required to degrade specific contaminants, such as chlorinated solvents or heavy metals. Bioaugmentation can reduce contaminant levels faster than natural attenuation or biostimulation. However, the introduced microorganisms must compete with native species and adapt to environmental conditions, potentially limiting their survival and effectiveness (Ławniczak et al., 2020).

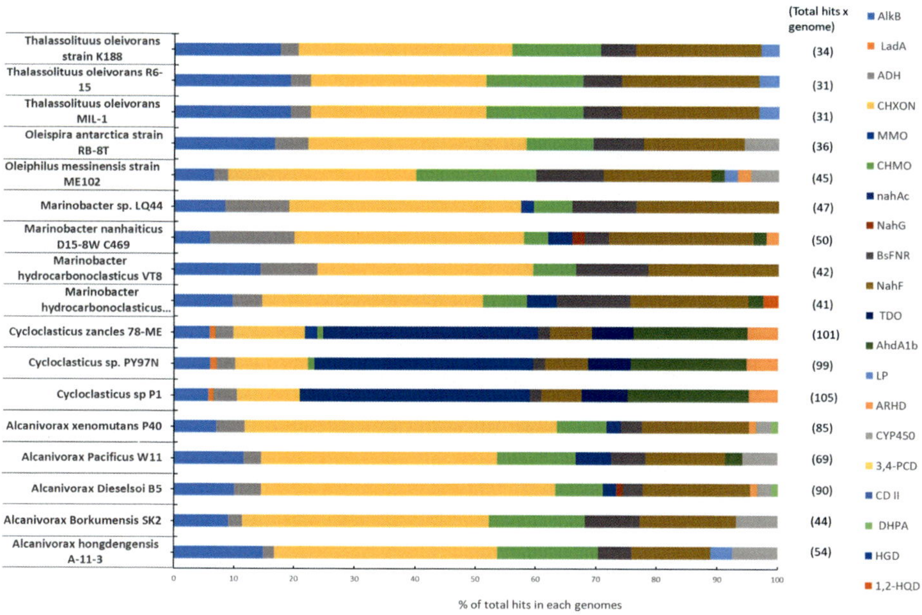

Figure 9.3. *General scheme of bioremediation strategies involving different microbial taxa.*

Source: Dell' Anno F, Rastelli E, Sansone C, Brunet C, Ianora A, Dell' Anno A. (2021). Bacteria, fungi and microalgae for the bioremediation of marine sediments contaminated by petroleum hydrocarbons in the omics era. Microorganisms, 9(8), 1695. https://doi.org/10.3390/microorganisms9081695.

9.2.4. Phytoremediation

Phytoremediation uses plants to clean up contaminated soils, sediments, and water. This strategy takes advantage of specific plants' natural abilities to absorb, accumulate, and detoxify pollutants through processes such as phytoextraction, phytostabilization, and rhizodegradation. Phytoremediation is particularly effective at removing heavy metals, pesticides, and organic compounds. The method is eco-friendly and can improve soil structure and fertility. However, it is slower than microbial bioremediation and is only applicable to shallow soils with low to moderate contamination levels (Malik and Goyal, 2022).

9.2.5. Co-Metabolism

Co-metabolism is the fortuitous transformation of contaminants by microorganisms as they degrade another substrate, which serves as their primary energy or carbon source. This strategy is critical for breaking down complex or recalcitrant pollutants that cannot be metabolized directly, such as certain chlorinated hydrocarbons. While effective, co-metabolism frequently necessitates the ongoing addition of the primary substrate, which can raise costs and complicate management (Verma and Kuila, 2019).

9.2.6. Bioventing

Bioventing is an in-situ remediation technique that improves aerobic biodegradation of organic contaminants in soil by introducing oxygen via direct air injection. By increasing oxygen availability, bioventing encourages native aerobic microorganisms to metabolize hydrocarbons and other organic pollutants. It works particularly well for treating petroleum hydrocarbons in unsaturated soils. Bioventing is relatively inexpensive and minimally invasive, but its effectiveness may be limited by soil permeability and contaminant distribution (Rajendran and Gunasekaran, 2019).

9.2.7. Biosorption

Biosorption is the passive uptake and binding of contaminants, specifically heavy metals, by biological materials such as microbial biomass, algae, and fungi. Various mechanisms drive this process, including ion exchange, complexation, and physical adsorption (Ławniczak et al., 2020). Biosorption is advantageous for treating wastewater and industrial effluents because of its high efficiency, low cost, and ability to recover and reuse biosorbent materials. However, biosorption is typically used as a preliminary treatment step and may need to be combined with other remediation methods to completely remove contaminants.

9.3. MICROBIAL CONSORTIA

Microbial consortia are complex communities of interacting microorganisms that live and function together in a given environment. These consortia are important for understanding microbial ecology because they demonstrate a level of organization and cooperation that exceeds the capabilities of individual microbial species. Microbial consortia research is important in fields such as bioremediation, wastewater treatment, and the human microbiome, where interactions between different microbial species are critical to ecosystem function and stability (Maier et al., 2009).

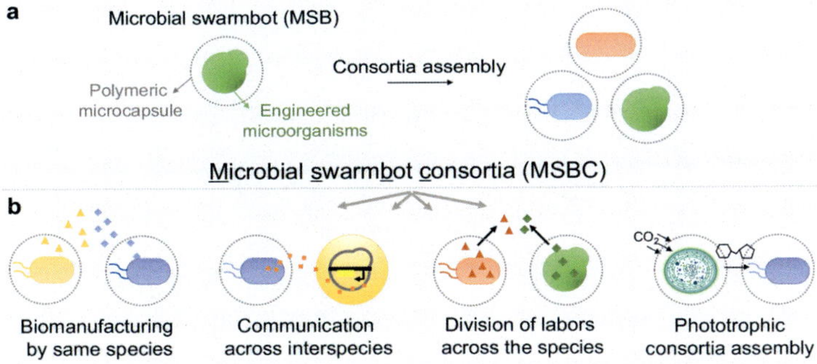

Figure 9.4. *Engineering consortia by polymeric microbial swarmbots.*

Source: Wang, L., Zhang, X., Tang, C. et al. (2022). Engineering consortia by polymeric microbial swarmbots. Nat. Commun., 13, 3879. https://doi.org/10.1038/s41467-022-31467-1.

9.3.1. Composition and Structure

Microbial consortia are made up of various species of bacteria, archaea, fungi, and, in some cases, algae and protozoa, all of which contribute to the community's overall functionality. The ecological niche they occupy often dictates the structure of these consortia, with factors such as nutrient availability, pH, temperature, and the presence of pollutants or toxins shaping species diversity. Microorganisms in these consortia interact in a variety of ways, including mutualism, commensalism, and competition, which affect the community's overall stability and productivity. The spatial organization of consortia, in which different microbial species occupy distinct niches or microenvironments, is also important for optimizing metabolic processes and reducing competitive interactions (Tan et al., 2023).

9.3.2. Functional Redundancy and Resilience

Functional redundancy is a key feature of microbial consortia, in which multiple species can perform similar metabolic functions. This redundancy increases the consortia's resilience, allowing them to maintain functional stability in the face of environmental fluctuations or perturbations. For example, in the degradation of complex organic matter in soil, different bacterial species may have overlapping enzymatic capabilities, ensuring that essential functions like cellulose degradation or nitrogen cycling continue even when specific species are lost or reduced in abundance (Malik and Goyal, 2022).

9.3.3. Synergy and Metabolic Cooperation

Microbial consortia demonstrate a high level of metabolic cooperation, with different species complementing each other's metabolic activities, resulting in improved overall performance. This synergy is most evident in syntrophic relationships, in which one species' metabolic byproducts serve as substrates for another. A classic example of such collaboration is the syntrophic interaction of hydrogen-producing bacteria and methanogenic archaea in anaerobic environments. The bacteria ferment organic compounds to produce hydrogen, which the methanogens use to produce methane, effectively lowering the partial pressure of hydrogen and propelling the fermentation process forward. This interdependence improves the efficiency of energy recovery in processes such as anaerobic digestion and is critical to the consortium's stability (Gautam et al., 2022; Malik and Goyal, 2022).

9.3.4. Applications in Bioremediation and Biotechnology

Microbial consortia are increasingly being harnessed in bioremediation and industrial biotechnology due to their ability to degrade complex pollutants and produce valuable biochemicals. In bioremediation, consortia are employed to degrade recalcitrant compounds, such as hydrocarbons, pesticides, and heavy metals, often achieving higher degradation rates and broader substrate specificity compared to individual species. The design and engineering of synthetic microbial consortia, where specific microbial species are selected and assembled based on their metabolic capabilities, is a growing area of research aimed at optimizing the degradation of pollutants or the production of biofuels and bioproducts (Waoo and Vyas, 2023).

9.4. BIOREMEDIATION TECHNOLOGIES

Bioremediation technologies utilize microbial metabolic capabilities to detoxify or degrade environmental pollutants. Microbial bioremediation employs a variety of technologies, which are broadly classified as in-situ or ex-situ, each tailored to specific contaminants and environmental conditions.

In-situ technology involves treating the contaminants in their natural settings while ex-situ technology involves transporting the contaminants to a different location for treatment. We will begin by describing some in-situ methods.

9.4.1. Biostimulation

Biostimulation is a popular bioremediation technique that improves the natural microbial degradation of contaminants by introducing nutrients, oxygen, or other electron acceptors into the contaminated environment. This method is especially useful in environments where nutrient deficiencies or insufficient electron acceptors limit the native microbial population. Biostimulation, which introduces essential nutrients such as nitrogen and phosphorus, can accelerate the microbial degradation of organic pollutants such as petroleum hydrocarbons and chlorinated solvents. Oxygen addition is also important, particularly in aerobic biostimulation, which increases the activity of aerobic microorganisms. The success of biostimulation relies on carefully balancing the added nutrients to prevent excessive microbial growth that could lead to the depletion of oxygen and other resources. This technique is commonly used in both in-situ and ex-situ bioremediation processes (Rajendran and Gunasekaran, 2019).

9.4.2. Bioaugmentation

Bioaugmentation is the deliberate introduction of specific microbial strains or consortia into a contaminated environment to improve pollutant degradation. This technology is especially effective when the native microbial community lacks the metabolic pathways required to degrade specific contaminants, such as polychlorinated biphenyls (PCBs), chlorinated solvents, or heavy metals. Bioaugmentation can be carried out using naturally occurring microbes that have been isolated and cultured or with genetically engineered organisms designed to degrade specific pollutants. The introduced microorganisms must be able to survive, compete with native species, and effectively degrade the target contaminants in the site's environment.

The success of bioaugmentation depends on the compatibility of the introduced strains with the existing microbial community and environmental factors such as pH, temperature, and the presence of co-contaminants (Sales Da Silva et al., 2020).

9.4.3. Biosparging

Biosparging is an in-situ bioremediation technique that involves injecting air or oxygen into the saturated zone of contaminated groundwater to boost the aerobic degradation of organic pollutants by indigenous microorganisms. This method is especially effective for the removal of petroleum hydrocarbons and other volatile organic compounds. Biosparging stimulates the metabolic activity of aerobic bacteria by increasing the oxygen concentration in groundwater, resulting in faster contaminant breakdown (Sales Da Silva et al., 2020; Tan et al., 2023). The technique is similar to bioventing, but is used in saturated subsurface environments rather than the vadose zone. The permeability of the soil, the depth of contamination, and the distribution of contaminants all influence biosparging effectiveness.

9.4.4. Bioventing

Bioventing is another in-situ bioremediation technology that promotes aerobic degradation of organic contaminants in the soil's unsaturated zone (vadose zone). Unlike biosparging, which injects air into the saturated zone, bioventing aims to increase oxygen availability in the vadose zone by injecting air or oxygen via wells. The added oxygen stimulates the activity of native aerobic microorganisms, which degrade hydrocarbons, solvents, and other organic pollutants. Bioventing is particularly effective for treating long-term, low-concentration contamination and is commonly used to remediate petroleum-contaminated soils. This method is advantageous because it is less disruptive and more cost-effective than other in-situ treatments; however, its success is dependent on soil permeability and the distribution of contaminants (Rajendran and Gunasekaran, 2019).

Some ex-situ methods include:

9.4.5. Composting

Composting is an ex-situ bioremediation technology that involves the controlled aerobic decomposition of organic contaminants in soils or solid waste materials. During composting, organic amendments such as manure, straw, or wood chips are mixed with contaminated soils to enhance microbial activity. The process relies on the natural microbial community to degrade complex organic pollutants like petroleum hydrocarbons, pesticides, and explosives into simpler, less toxic substances. The addition of organic matter improves soil structure, aeration, and moisture retention, resulting in ideal conditions for microbial degradation. Composting is particularly effective for treating large volumes of contaminated soil; however, it requires careful monitoring of temperature, moisture, and aeration to ensure that microbial activity is maintained at optimal levels (Sales Da Silva et al., 2020; Malik and Goyal, 2022).

9.4.6. Landfarming

Another type of ex-situ bioremediation is landfarming, which involves spreading contaminated soils over a large surface area and tilling them on a regular basis to promote aerobic microbial degradation. Organic amendments, nutrients, and, on occasion, moisture are used to boost microbial activity and contaminant degradation. Landfarming is a popular method for treating petroleum hydrocarbons and other organic contaminants in the oil and gas industry. Periodic tilling aerates the soil, ensuring that there is enough oxygen to support aerobic degradation. Landfarming is a cost-effective and simple method, but it requires large land areas and can take several months to achieve significant contaminant reduction (Waoo and Vyas, 2023).

9.4.7. Bioreactors

Bioreactors are engineered systems that treat contaminated media like soil, sediment, or wastewater under controlled conditions to maximize microbial activity. These systems can be designed for aerobic or anaerobic conditions, depending on the contaminants and the desired degradation pathways. Bioreactors have several advantages, including the ability to precisely control environmental parameters such as temperature, pH, oxygen levels, and nutrient concentrations, which can significantly improve the efficiency of biodegradation. They are especially effective in the treatment of highly contaminated or hazardous materials that require extensive treatment. However, bioreactors are generally more expensive and difficult to operate than in-situ technologies, and they are typically used for smaller amounts of contaminated material (Rajendran and Gunasekaran, 2019; Tan et al., 2023).

9.5. BIOREMEDIATION SUCCESS STORIES

Bioremediation has been successfully applied in numerous real-world scenarios to address significant environmental challenges. Below are some notable examples where bioremediation has played a crucial role in mitigating environmental contamination:

9.5.1. Exxon Valdez Oil Spill, Alaska (1989)

The Exxon Valdez oil spill in 1989 was one of the worst environmental disasters in history. The Exxon Valdez tanker collided with a reef in Prince William Sound, Alaska, spilling approximately 11 million gallons of crude oil into the pristine waters. The oil contaminated 1,300 miles of coastline, harming marine life and local ecosystems. Bioremediation played an important role in the cleanup efforts. The primary approach was biostimulation, which involved applying nitrogen and phosphorus fertilizers to the affected shorelines to stimulate the growth of indigenous oil-degrading bacteria. These nutrients increased microbial activity, hastening the natural degradation of the oil. Studies have shown that biostimulation significantly increased the rate of oil biodegradation, leading to a faster recovery of the contaminated beaches (Hazen et al., 2010). This event marked one of the first large-scale applications of bioremediation in response to an environmental disaster.

9.5.2. Bunker C Oil Spill, Buzzards Bay, Massachusetts (2003)

In 2003, the Bouchard Barge 120 dumped about 98,000 gallons of Bunker C fuel oil into Buzzards Bay, Massachusetts. The heavy oil spread quickly, polluting the shoreline and endangering local wildlife. Bioremediation was a crucial part of the cleanup strategy. Nutrient amendments were used to stimulate indigenous microbial communities that are capable of degrading heavy oil fractions. The bioremediation efforts focused on improving aerobic degradation processes, which are more effective at breaking down the complex hydrocarbons found in Bunker C oil. The significant reduction in oil residues in the affected areas over time demonstrates the success of these efforts. This case demonstrated the effectiveness of bioremediation in dealing with oil spills that contained heavier, more complex hydrocarbons.

9.5.3. Ogoni Land Oil Spill, Nigeria

The Ogoni Land in Nigeria's Niger Delta region has been heavily polluted as a result of decades of oil extraction activities. The region has seen numerous oil spills, resulting in severe environmental degradation and health risks for local residents. In response to this crisis, bioremediation has been proposed as part of the cleanup strategy. The United Nations Environment Program (UNEP) emphasized the potential of bioremediation, specifically bioaugmentation and biostimulation, for restoring contaminated soils and water bodies. These techniques aim to increase the activity of indigenous and introduced microorganisms capable of degrading hydrocarbons found in oil-polluted environments (Reddy et al., 2012). While bioremediation efforts are ongoing, they are an important part of the larger environmental restoration efforts in Ogoni Land.

9.5.4. Deepwater Horizon Oil Spill, Gulf of Mexico (2010)

The Deepwater Horizon oil spill in 2010 spilled approximately 210 million gallons of crude oil into the Gulf of Mexico, making it the largest marine oil spill in history. Bioremediation was an important part of the response strategy, particularly the use of dispersants to increase the oil's bioavailability to native hydrocarbon-degrading bacteria. Although controversial, the use of dispersants facilitated the formation of oil droplets that were easier for microorganisms to degrade. Studies revealed that naturally occurring bacteria, such as those in the genus Alcanivorax, played an important role in breaking down the oil in the deep-sea environment (Hazen et al., 2010). The bioremediation efforts contributed to the significant reduction of oil in the water column, highlighting the effectiveness of microbial degradation in mitigating large-scale oil spills.

9.5.5. Hudson River PCBs Remediation, New York

Between the 1940s and 1970s, General Electric's industrial discharges heavily contaminated the Hudson River in New York with PCBs. PCBs are persistent organic pollutants that endanger human health and the environment. The Hudson

River remediation strategy included bioremediation as a supplement to dredging. Bioremediation efforts aimed to improve the natural degradation of PCBs by indigenous microorganisms, particularly through the use of bioaugmentation and biostimulation techniques (Reddy et al., 2012). While the primary method involved physically removing contaminated sediments, bioremediation has helped to reduce PCB levels in less accessible areas of the river.

9.5.6. Pesticide Degradation in Agricultural Soils, Canada

Pesticide use in agriculture in Canada has resulted in the accumulation of toxic residues in soils, which endangers both the environment and human health. Bioremediation has been used successfully to degrade pesticide residues, particularly with microbial consortia that have the necessary enzymatic pathways to break down complex pesticide molecules. For example, introducing specific bacteria that can metabolize atrazine, a commonly used herbicide, has resulted in its degradation (Hazen et al., 2010). These bioremediation efforts not only reduced pesticide levels in agricultural soils but also helped to restore soil health, making crop production safer.

9.6. BIOREMEDIATION CHALLENGES

Bioremediation is hampered by several challenges that limit its widespread application. These challenges include the bioavailability of contaminants, the complexity of contaminant mixtures, environmental conditions, the adaptation and survival of microorganisms, the formation of toxic metabolites, and economic and regulatory barriers. Addressing these issues is crucial to enhancing the efficiency and feasibility of bioremediation strategies.

9.6.1. Bioavailability of Contaminants

The bioavailability of pollutants is an important aspect in determining the success of bioremediation operations. Bioavailability is the degree to which pollutants are available to microorganisms for breakdown. Many pollutants are hydrophobic and adsorb heavily to soil particles, limiting their availability to microorganisms. This limited availability can greatly impede the breakdown process, making bioremediation less effective. Studies have demonstrated that the physical and chemical features of pollutants, as well as the qualities of the soil or sediment, all have an important influence in influencing bioavailability. Contaminants like PAHs are renowned for their limited bioavailability due to their great affinity for organic materials in soils (Alexander, 2000). To overcome this challenge, strategies such as the use of surfactants or biosurfactants have been proposed to enhance the solubility and desorption of hydrophobic contaminants, thereby increasing their bioavailability to degrading microorganisms (Denecke et al., 2018).

9.6.2. Complex Contaminant Mixtures

Pollutants are rarely found in isolated forms in many contaminated environments; instead, they are frequently a part of complex mixtures. These combinations may include both organic and inorganic substances, which could complicate the bioremediation procedure due to potential interactions. Certain contaminants can impede the breakdown of other contaminants, or the accumulation of several contaminants can overburden microbial communities, resulting in partial degradation. For example, it can be especially problematic when heavy metals and organic pollutants coexist because heavy metals can hinder the activity of microbes that are necessary for the breakdown of organic pollutants. Furthermore, distinct pollutants' degradation processes can interact with one another, resulting in the buildup of partially degraded compounds. To address these difficulties, a thorough understanding of the interactions between contaminants is required, as is the development of bioremediation techniques capable of targeting many contaminants at the same time (Maier et al., 2009; Malik and Goyal, 2022).

9.6.3. Environmental Conditions

The efficiency of bioremediation is greatly impacted by environmental factors such as temperature, pH, oxygen availability, and nutrient levels. Due to their sensitivity to variations in these conditions, the microorganisms participating in bioremediation can have their activity enhanced or inhibited. For example, the enzymatic activity that propels the biodegradation process requires ideal temperatures. According to Gentry et al. (2004), temperature extremes can kill out entire microbial communities or drastically slow down their metabolism. In a similar vein, the lifespan of microorganisms and the solubility of pollutants can be affected by pH levels. For example, extremes in acidity or alkalinity can be harmful to microbiological life, which reduces the efficacy of bioremediation (Malik and Goyal, 2022). Oxygen availability is also crucial for aerobic degradation processes, and its limitation can slow down the breakdown of contaminants that require oxygen. Managing these environmental conditions is therefore vital for the success of bioremediation efforts.

9.6.4. Adaptation and Survival of Microorganisms

The ability of microorganisms to adapt to and survive in contaminated settings is essential to the effectiveness of bioremediation. To efficiently digest pollutants, microorganisms need to not only live in hostile environments but also continue with their metabolic processes. Nevertheless, it is frequently the case that native microbial populations are either insufficiently abundant or lack the metabolic pathways required to break down specific pollutants. It may be possible to introduce foreign microorganisms with particular degradative properties, although these introduced species frequently encounter difficulties adjusting to their new surroundings and contending with native species (Maier et al., 2009). Therefore, microbial communities can be adversely affected by toxic contaminants, leading to a decrease in microbial diversity and activity. Ensuring the survival and activity of microorganisms requires careful consideration of the microbial ecology and the development of strategies to enhance microbial adaptation and resilience.

9.6.5. Formation of Toxic Metabolites

During the biodegradation process, pollutants are converted into intermediate metabolites, which can be more dangerous than the parent chemicals. The development of these hazardous metabolites is a serious barrier to bioremediation since it might lead to secondary contamination. For example, the breakdown of chlorinated solvents can produce vinyl chloride, a very poisonous and carcinogenic chemical (Rajendran and Gunasekaran, 2019). The accumulation of such metabolites can reduce the overall efficacy of bioremediation while also posing additional dangers to human health and the environment. As a result, it is critical to continuously monitor the degradation process and guarantee that the bioremediation technologies used are capable of fully mineralizing pollutants into safe end products.

9.6.6. Economic and Regulatory Barriers

Economic and regulatory hurdles also contribute significantly to the limited use of bioremediation. Bioremediation costs can be substantial, especially for large-scale or complex sites that require long-term treatment. Furthermore, the economic viability of bioremediation is frequently undermined by the availability of less ecologically friendly alternatives, such as chemical treatments. Regulatory frameworks can also be challenging, as bioremediation projects must adhere to environmental standards that may differ among locations. Obtaining permissions and following regulatory standards can be time-consuming and expensive. Furthermore, public image and acceptance of bioremediation might have an impact on regulatory choices, especially if the process's safety and effectiveness are questioned. To address these economic and regulatory issues, scientists, legislators, and industry stakeholders must work together to create cost-effective and long-term bioremediation techniques (Malik and Goyal, 2022).

9.7. BIOREMEDIATION REGULATIONS

Bioremediation implementation is subject to a complicated web of rules that ensure safety, efficacy, and environmental protection. These restrictions varied greatly among jurisdictions, reflecting the various legal, environmental, and societal settings in which bioremediation is used. Understanding these restrictions is critical for academics, practitioners, and politicians participating in bioremediation initiatives.

9.7.1. Environmental Protection Agency (EPA) Regulations (U.S.)

In the United States, the EPA regulates bioremediation activities. The EPA's power stems primarily from several significant pieces of legislation, including the Comprehensive Environmental Response, Compensation, and Liability Act (CERCLA), sometimes known as Superfund, and the Resource Conservation and Recovery Act (RCRA). Under CERCLA, the EPA is in charge of the cleanup of hazardous waste sites, many of which use bioremediation as a remediation approach. The EPA establishes rules for the appropriate use of bioremediation, ensuring that it does not worsen environmental or health hazards.

The EPA's regulations are aimed at examining the risks connected with bioremediation procedures, namely the use of genetically modified organisms (GMOs) and other bioengineered goods. The Toxic Substances Control Act (TSCA) also has an impact, requiring that all GMOs used in bioremediation be thoroughly examined and evaluated for safety before deployment. Furthermore, the EPA requires comprehensive monitoring and reporting on bioremediation programs to verify compliance with environmental requirements and to assess the efficacy of remedial activities (Maier et al., 2009).

9.7.2. State Regulations

States in the United States have their own laws controlling bioremediation in addition to federal regulation overseen by the EPA. Although different states have different environmental concerns and policy agendas, these regulations might range greatly from one another. For example, states with a history of industrial contamination, like New York and California, have created extensive regulatory frameworks to oversee bioremediation initiatives. In comparison to federal standards, these frameworks frequently include stricter environmental impact studies and permitting procedures.

State laws may also cover particular pollutants or ecosystems that are especially important to the state. States that experience a lot of agricultural activity, for instance, can have particular regulations for bioremediation of pesticide-contaminated soils, whereas states that are coastline would concentrate on bioremediation in marine ecosystems. Although state agencies and the EPA frequently collaborate, they are free to impose further rules or limitations as necessary (Maier et al., 2009).

9.7.3. International Regulations

Bioremediation regulations are determined by a combination of international environmental accords, regional policies, and national legislation. The European Union (EU), for example, has created a thorough legislative framework called the Environmental Liability Directive (ELD), which attempts to prevent and repair environmental damage. The EU's approach to bioremediation stresses the precautionary principle, which necessitates extensive risk evaluations and strict adherence to environmental standards.

Countries such as Canada and Australia have also adopted national bioremediation legislation, which frequently incorporates international principles such as those issued by the International Organization for Standardization (ISO). Regulatory frameworks for bioremediation in underdeveloped countries are continuously changing, influenced by international aid programs and collaboration with wealthier countries (Hewitt, 2004).

9.7.4. Occupational Safety and Health Administration (OSHA) Regulations

The OSHA oversees the safety and health elements of bioremediation initiatives, especially those involving hazardous materials or settings. OSHA standards are intended to protect bioremediation workers from potential risks such as hazardous chemical exposure, biological hazards, and unsafe working conditions. Employers must comply with these rules by implementing safety measures, providing adequate protective equipment, and conducting frequent worker training.

OSHA's Hazardous Waste Operations and Emergency Response (HAZWOPER) rules are especially pertinent to bioremediation since they establish precise procedures for handling and remediating hazardous waste. Compliance with OSHA rules is critical not only for worker safety but also for the legal and financial viability of bioremediation efforts (Hewitt, 2004).

9.7.5. Research and Innovation

Innovation and research are essential to bioremediation because they drive the creation of new methods and increase the efficacy of those that already exist. Regulatory agencies such as the European Chemicals Agency (ECHA) and the

EPA are crucial in monitoring bioremediation research endeavors, especially when new technologies like nanotechnology and synthetic biology are involved.

Before being extensively used, new bioremediation systems are frequently subject to regulations requiring extensive testing and validation. This entails evaluating possible effects on the environment and human health in addition to making sure the technologies abide by current safety and environmental regulations. Regulatory frameworks also encourage innovation by providing guidelines for conducting research safely and responsibly, balancing the need for environmental protection with the potential benefits of new bioremediation approaches (Lederberg and Relman, 2009).

9.8. BIOREMEDIATION MONITORING AND ASSESSMENT

Bioremediation monitoring and assessment are critical components in evaluating the effectiveness of bioremediation strategies for environmental cleanup. These processes ensure that the remediation activities achieve their intended goals, maintain environmental safety, and provide data for optimizing treatment protocols. Monitoring involves the ongoing collection of data during the bioremediation process, while assessment focuses on evaluating the overall performance and efficacy of the remediation efforts.

9.8.1. Monitoring

For bioremediation to be safe and effective, monitoring must take into account a number of important factors.

- **Contaminant Concentration**: Tracking the amount of pollutants present is essential for evaluating how well bioremediation is going. To determine the concentrations of certain pollutants, such as hydrocarbons, heavy metals, or solvents, routine sampling and analysis of soil, water, or air is carried out. Accurate and sensitive measurements can be obtained by using methods like high-performance liquid chromatography (HPLC) and gas chromatography-mass spectrometry (GC-MS).
- **Site-specific Parameters**: The bioremediation process might be influenced by the particular features of each remediation site. It's critical to keep an eye on site-specific factors including temperature, pH, moisture content, and soil texture. Microbial activity and pollutant bioavailability are impacted by these variables. For example, the solubility of pollutants and the activity of degrading microorganisms can be affected by the pH of the soil (Khan et al., 2021).

- **Microbial Activity**: Determining how well microorganisms are breaking down pollutants requires evaluating microbial activity. Evaluation of the microbial community's health and activity is aided by techniques including microbial biomass measurement, enzyme activity assays, and molecular approaches like quantitative PCR (qPCR). This facilitates the identification of the microorganisms present and the relationship between their activity and the decline in pollutant levels.
- **Environmental Conditions**: Several factors, including temperature, oxygen saturation, and nutrient availability, can have a big impact on how effective bioremediation is. By changing the operational parameters to promote microbial activity and pollutant breakdown, monitoring these variables aids in the optimization of the bioremediation process. For instance, the availability of nutrients is necessary to maintain microbial development, and oxygen levels are crucial in aerobic bioremediation (Verma and Kuila, 2019).

9.8.2. Assessment

Assessment involves evaluating the effectiveness of bioremediation efforts based on collected data and comparing it with predetermined goals.

- **Performance Evaluation**: This evaluation determines if the goals of the bioremediation process are being met. This entails assessing whether the cleanup objectives are being fulfilled and comparing the decrease in pollutant concentrations to anticipated results. Key criteria in this assessment include performance indicators including microbial community health and pollutant reduction rates.
- **Remediation Goals**: A successful evaluation depends on the establishment of precise remediation goals. Usually, these objectives include precise goals for reducing contaminants and restoring the site's conditions. The objectives must be founded on site-specific features, risk evaluations, and regulatory requirements. The success of the cleanup can be gauged by how well these objectives are met.
- **Data Analysis and Reporting**: To make inferences about the bioremediation process, data analysis entails evaluating the gathered monitoring data. To assess trends, spot anomalies, and forecast performance in the future, statistical techniques and modeling can be employed. Reporting includes recording the results and offering suggestions for additional work or modifying the repair plan (De Mandal et al., 2022).

9.8.3. Challenges and Considerations

Bioremediation monitoring and assessment face several challenges and considerations that must be addressed to ensure successful remediation.

- **Complexity of Environmental Systems**: Environmental systems are inherently complex, and variables such as diverse pollutant distribution, changeable site circumstances, and unpredictable microbial behavior can impede monitoring and evaluation efforts.
- **Data Interpretation**: Proper interpretation of monitoring data necessitates a full understanding of the bioremediation process and environmental dynamics. Misinterpretation of data can lead to inaccurate assumptions about the efficiency of remedial efforts.
- **Regulation and Stakeholder Requirements**: Compliance with regulatory standards, as well as resolving stakeholder concerns, can make monitoring and assessment processes more complex. It is critical to match remediation efforts with regulatory requirements and communicate clearly with stakeholders.

As a result, bioremediation monitoring and assessment are critical to assuring the success of environmental cleaning initiatives. Bioremediation efficiency can be precisely evaluated and optimized by closely monitoring contaminants, site-specific circumstances, microbial activity, and environmental parameters, as well as rigorously assessing performance, goals, and data. Addressing obstacles and considerations ensures that remediation efforts are effective and in line with regulatory standards (Verma and Kuila, 2019).

9.9. BIOREMEDIATION FUTURE TRENDS

The field of bioremediation is rapidly evolving, driven by technological advancements and a growing understanding of environmental systems. Future trends are poised to significantly enhance the efficacy and scope of bioremediation practices. Key areas of development include synthetic biology and genetic engineering, advancements in omics technologies, applications of nanotechnology, refined evaluation protocols, and strategies to adapt to climate change. These trends represent a leap towards more efficient, targeted, and sustainable environmental cleanup solutions.

9.9.1. Synthetic Biology and Genetic Engineering

Synthetic biology and genetic engineering are revolutionizing bioremediation by enabling the design and creation of microorganisms with enhanced capabilities for pollutant degradation.

Engineered Microbes: Synthetic biology enables the exact manipulation of microbial genomes to improve their ability to break down specific pollutants. Engineered bacteria, for example, can be programmed to degrade complex hydrocarbons or heavy metals more effectively than natural strains. This personalization can be achieved by introducing new metabolic pathways or improving current ones.

Gene Editing Technologies: Advanced gene editing techniques like CRISPR-Cas9 provide precise control over genetic alterations in microorganisms. This method enables the creation of microorganisms with optimal degradation pathways and greater resistance to environmental stressors (Gibson et al., 2019). Gene editing also enables the formation of synthetic consortia in which multiple modified strains collaborate synergistically to combat diverse or persistent contaminants (Gibson et al., 2019).

9.9.2. Advancement in Omics Technologies

Omics technologies, encompassing genomics, proteomics, and metabolomics, are providing deeper insights into microbial functions and interactions during bioremediation.

- **Metagenomics:** This method uses ambient DNA to detect and characterize microbial populations without the requirement for cultivation. Metagenomics allows researchers to identify microbial species and genes involved in pollutant degradation, providing useful information for enhancing bioremediation procedures (Denecke et al., 2018).
- **Proteomics and Metabolism:** Proteomics investigates proteins and their roles, whereas metabolomics studies metabolic products. Together, these technologies provide a comprehensive perspective of the microbial activity and metabolic pathways involved in bioremediation. Understanding protein expression patterns and metabolic fluxes allows researchers to improve the design and implementation of microbial treatments.
- **Systems Biology:** Integrating data from genomes, proteomics, and metabolomics enables the modeling of complex microbial interactions and their consequences on bioremediation processes. This comprehensive method aids in predicting microbial responses to various pollutants and environmental situations, resulting in more effective remedial efforts (Denecke et al., 2018).

9.9.3. Nanotechnology Applications

Nanotechnology is emerging as a potent tool for improving bioremediation processes using novel techniques.

- **Nanoparticle Utilization:** Nanoparticles can be employed to increase the bioavailability of pollutants or to better distribute remedial agents. Engineered nanoparticles, for example, can help reduce pollutants like heavy metals and organic solvents by providing reactive surfaces and promoting microbial activity (Malik and Goyal, 2022).
- **Nano-Enhanced Bioreactors:** Introducing nanotechnology into bioreactors can improve the effectiveness of bioremediation processes. Nanomaterials can help bioreactors work better by raising response rates, improving pollutant removal, and monitoring real-time changes in bioreactor conditions (Malik and Goyal, 2022).

9.9.4. Evaluation Protocols

Refined evaluation techniques are crucial for determining the efficacy of bioremediation operations and ensuring that remediation objectives are attained.

- **Performance Measures:** Advanced measures are being created to assess the effectiveness of bioremediation methods. These include not only reducing pollutant concentrations but also improving microbial community health and system performance. Comprehensive performance assessments paint a more realistic picture of remediation effectiveness and areas for development (Denecke et al., 2018).
- **Real-Time Monitoring:** The use of real-time sensors and data analytics allows for continuous monitoring of bioremediation processes and ambient conditions. This feature enables fast alterations to remediation tactics based on current data, increasing the overall efficiency and responsiveness of bioremediation activities (Cheng et al., 2021).

9.9.5. Climate Change Adaptation

Adapting bioremediation strategies to the impacts of climate change is becoming increasingly important.

- **Resilient Strategies:** Climate change can affect environmental factors such as temperature, precipitation, and soil qualities. Developing bioremediation technologies that are resistant to these changes is critical to sustaining efficacy. The research focuses on identifying and developing microorganisms that can flourish in a variety of climates and continue to operate well in the face of environmental challenges (De Mandal et al., 2022).

- **Integrated Approaches:** Addressing climate change necessitates combining bioremediation with larger environmental management techniques. This includes examining how climate change may affect pollutant behavior and cleanup performance, as well as establishing adaptive management plans that take these effects into account (Alonso, 2016).

Developments in synthetic biology, omics technologies, nanotechnology, assessment methodologies, and climate change adaptability are expected to influence the direction of bioremediation in the future. These developments should improve bioremediation initiatives' efficiency, accuracy, and sustainability while bringing more creativity and efficiency to the handling of challenging environmental issues.

CHAPTER 10

MICROBIAL ETHICS AND REGULATIONS

Contents

Microbial ethics and regulations are critical in ensuring the responsible conduct of microorganism-related research and practice. As biotechnological applications become more complex, the ethical implications of microbial research have received increased attention. Microorganisms, despite their small size, have enormous potential in a variety of fields, including medicine, agriculture, and environmental management. However, their manipulation and use raise serious ethical concerns, particularly regarding biosafety, biosecurity, and the possibility of unintended consequences (Woodworth et al., 2017).

Ethical considerations in microbial research include the safe handling and containment of pathogenic microorganisms to avoid accidental release, the responsible use of genetic modification techniques, and an equitable distribution of benefits from microbial discoveries. Furthermore, the dual-use nature of microbial research, in which scientific advances may be misapplied for harmful purposes, highlights the importance of stringent regulatory frameworks (Desikan et al., 2011). Regulations at both the national and international levels seek to address these ethical issues by establishing laboratory practice standards, encouraging research transparency, and ensuring biosafety protocol compliance. The convergence of ethics and regulation in microbial research protects public health and the environment while also encouraging responsible scientific innovation (Desikan et al., 2011).

10.1. ETHICAL CONSIDERATIONS IN MICROBIOLOGY

In microbiology, several ethical considerations are critical to ensuring that research and practice are carried out responsibly and with regard to human, environmental, and societal values. Important ethical considerations include:

10.1.1. Biosecurity

Biosecurity is an important ethical consideration, especially given the dual-use nature of many microbiological technologies. Dual-use research is scientific work that, while intended for beneficial purposes, has the potential to be abused in ways that could cause harm, such as the development of biological weapons (Desikan et al., 2011). Ethical biosecurity practices include enforcing strict controls to prevent unauthorized access to dangerous pathogens and ensuring that research does not contribute to the spread of biological weapons. This includes the creation and implementation of policies governing the handling, storage, and distribution of biological materials. Researchers must also engage in ongoing risk assessment and management to identify and mitigate potential threats to their work (Woodworth et al., 2017).

10.1.2. Biosafety

Biosafety is closely related to biosecurity, but it focuses on the safe handling and containment of microorganisms to protect laboratory personnel, the general public, and the environment from potentially hazardous agents. Biosafety practices are based on principles such as containment, risk assessment, and the use of appropriate protective equipment. Biosafety levels (BSL) range from BSL-1 (low risk) to BSL-4 (high risk), with corresponding containment practices and protocols. Ethically, researchers must strictly adhere to these guidelines to avoid accidental pathogen releases that could result in infectious disease outbreaks (Desikan et al., 2011). The ethical imperative of biosafety also includes the responsibility of adequately training laboratory personnel in safe practices and emergency protocols.

10.1.3. Informed Consent

Informed consent is an essential component of ethical research, especially in clinical microbiology, where human participants are involved. Before agreeing to participate, participants must be completely informed about the nature, purpose, risks, and potential benefits of the research. Informed consent is more than a formality; it embodies the ethical principle of respect for autonomy, ensuring that people make voluntary and informed decisions about their participation in research. Researchers must communicate information clearly and without coercion, ensuring that participants fully understand the implications of their participation. In cases involving vulnerable populations, additional safeguards are required to ensure that consent is truly informed and voluntary (Woodworth et al., 2017).

10.1.4. Privacy and Confidentiality

Privacy and confidentiality are critical ethical concerns, given the growing use of genetic and genomic data in research. Researchers have an ethical obligation to protect the privacy of individuals whose biological samples are studied, ensuring that personal information is kept confidential and not disclosed without their consent. This includes implementing strong data security measures to prevent unauthorized access and ensuring that data sharing practices adhere to ethical and legal standards. The ethical obligation to maintain confidentiality is especially important in infectious disease research, where breaches could result in stigma or discrimination against individuals or communities (Woodworth et al., 2017).

10.1.5. Equitable Access and Benefit Sharing

Equitable access and benefit-sharing are critical ethical concerns, especially in global health and environmental microbiology. The Nagoya Protocol

(2010) emphasizes the equitable distribution of benefits derived from the use of genetic resources. This is especially important when conducting research using biological resources from low- and middle-income countries. Ethical practice requires that these countries benefit from the research, whether through capacity development, technology transfer, or a share of the profits from commercialization. Researchers must also ensure that local communities are not exploited and that their contributions are properly recognized and compensated (Meechan and Potts, 2020).

10.1.6. Environmental Impact

The potential environmental impact of microbiology research, especially synthetic biology and the use of genetically modified organisms (GMOs), raises serious ethical concerns. Researchers must conduct extensive environmental risk assessments to determine the potential consequences of releasing GMOs or other modified microorganisms into the environment. Ethical stewardship entails minimizing ecological disruptions while avoiding unintended consequences such as the spread of antibiotic resistance or the disruption of local ecosystems (Meechan and Potts, 2020). This includes following environmental regulations, conducting regular monitoring, and being open about potential risks.

10.1.7. Animal Welfare

Animal welfare is an important ethical consideration in microbiology, particularly in studies that use animal infection models. Researchers have an ethical obligation to follow the 3Rs principles—Replacement, Reduction, and Refinement—to minimize animal use and suffering. This includes replacing animal models with alternatives whenever possible, reducing the number of animals used, and improving experimental techniques to reduce pain and suffering. Ethical considerations include the humane treatment of animals throughout the research process, such as housing, care, and euthanasia (Desikan et al., 2011; Meechan and Potts, 2020).

10.2. BIOSAFETY AND BIOSECURITY

Biosafety and biosecurity are two interrelated ethical considerations that are critical to ensuring the responsible conduct of microorganism-based research and practice. Both concepts are critical to protecting public health, environmental integrity, and global security by reducing the risks associated with the handling and manipulation of potentially dangerous biological agents. We will discuss the ethical dimensions of biosafety and biosecurity, emphasizing their importance, challenges, and frameworks for implementation in microbiological research (Woodworth et al., 2017).

10.2.1. Biosafety

The fundamental goal of biosafety is to protect laboratory personnel, the general public, and the environment from infectious agents and biohazards. The ethical imperative of biosafety stems from the need to prevent harm and ensure that microbiological research does not unintentionally result in outbreaks of infectious diseases or environmental contamination. This consideration is especially important in research involving pathogenic microorganisms, recombinant DNA technology, and synthetic biology, where the risk of harm is high (Desikan et al., 2011).

The ethical foundation of biosafety is based on the principle of nonmaleficence, which requires that actions do not cause harm. In microbiology, this translates into stringent risk assessment and management practices aimed at reducing the likelihood of hazardous agent exposure or release by accident. Microbiological research laboratories are classified into BSLs, which range from BSL-1 (low risk) to BSL-4 (high risk), with each level requiring specific containment practices and protocols to ensure safety. Adherence to these protocols is more than a regulatory requirement; it is an ethical obligation to protect human health and the environment (Stasi and Thongpravati, 2024).

One of the most important ethical challenges in biosafety is ensuring that laboratory personnel are adequately trained and equipped to handle biohazards safely. This includes not only technical training but also fostering a culture of safety in which ethical considerations are built into daily laboratory procedures. Researchers must be aware of the risks associated with their work and take proactive steps to mitigate them (Woodworth et al., 2017). This includes wearing appropriate personal protective equipment (PPE), maintaining a secure laboratory environment, and adhering to established protocols for handling, storing, and disposing of biohazards.

Additionally, the ethical responsibility of biosafety goes beyond the laboratory. Researchers must consider the larger implications of their work, including the possibility of environmental contamination. For example, releasing GMOs or synthetic microorganisms into the environment may have unintended ecological consequences, such as disrupting local ecosystems or contributing to the spread of antibiotic resistance. Ethical biosafety practices necessitate thorough environmental risk assessments and the implementation of containment strategies to avoid such outcomes.

10.2.2. Biosecurity

Biosecurity, while closely related to biosafety, is concerned with preventing the misuse of microbiological research for harmful purposes such as bioterrorism or the development of biological weapons. Biosecurity ethics is based on the principle of beneficence, which requires researchers to ensure that their work benefits the public good while causing no harm. In microbiology, this entails preventing potentially dangerous biological materials and information from falling into the wrong hands (Meechan and Potts, 2020).

The dual-use nature of microbiological research raises serious ethical concerns in biosecurity. Many of the techniques and knowledge developed in microbiology have legitimate scientific and medical applications, but they can also be used to cause harm. For example, studies on highly pathogenic viruses, such as the H5N1 influenza virus, have the potential to improve our understanding of viral transmission and inform public health responses. However, if such research is not properly safeguarded, it may be used to create more virulent strains of the virus, posing a serious threat to global security (Baltz et al., 2010; Desikan et al., 2011).

Ethical biosecurity practices include creating and enforcing policies and procedures to prevent unauthorized access, use, or dissemination of dangerous biological agents and information. This includes both physical security measures, such as restricting access to laboratories, and cyber security measures, which protect sensitive data. Furthermore, ethical biosecurity requires researchers to conduct risk-benefit analyses to carefully consider the potential consequences of their work and to implement safeguards to reduce the risk of misuse (Desikan et al., 2011).

Transparency and accountability are key ethical responsibilities in biosecurity. Researchers must be open about the risks associated with their work and engage with regulatory bodies, policymakers, and the general public to ensure that biosecurity concerns are addressed adequately. This includes reporting any security breaches or incidents that could jeopardize biosecurity, as well as contributing to the development of microbiology biosecurity guidelines and best practices (Meechan and Potts, 2020).

The global scope of microbiological research adds to the ethical challenges of biosecurity. Pathogens do not respect borders, and using biological agents in one country can have far-reaching consequences. As a result, international cooperation and the harmonization of biosecurity standards are critical to addressing the ethical issues raised by microbiology's dual-use nature. Researchers, institutions, and governments must collaborate to create and implement biosecurity measures that are effective in diverse cultural, legal, and regulatory contexts (Woodworth et al., 2017).

10.2.3. Integrating Biosafety and Biosecurity

Biosafety and biosecurity are distinct concepts, but they are inextricably linked and should be incorporated into a comprehensive ethical approach to microbiological research. Both considerations emphasize the importance of preventing harm, whether from accidental exposure or intentional misuse of biological agents. To accomplish this, researchers must implement a dual framework that includes both biosafety and biosecurity in their daily practices (Baltz et al., 2010).

This integrated approach necessitates ongoing education and training to ensure that researchers understand the ethical implications of their work and are prepared to face the challenges of both biosafety and biosecurity. Furthermore, institutions must foster a culture of accountability in which ethical considerations are integrated into research planning, execution, and dissemination (Meechan and Potts, 2020).

10.3. REGULATION OF MICROBIAL RESEARCH

Microbial research is the foundation of modern science, enabling advances in medicine, agriculture, environmental management, and biotechnology. However, due to the potential risks associated with the handling and manipulation of microorganisms, including pathogenic species and GMOs, microbial research is subject to strict regulatory frameworks. These regulations protect public health, ensure environmental safety, and prevent the misuse of scientific knowledge.

10.3.1. International Treaties and Conventions

Several treaties and conventions govern microbial research on a global scale, establishing global standards for biosafety, biosecurity, and ethical practice. The Convention on Biological Diversity (CBD) is one of the most important of these, emphasizing biological diversity conservation, the sustainable use of its components, and the fair and equitable distribution of benefits from genetic resources. The Nagoya Protocol, an extension of the CBD, specifically addresses access to genetic resources and the sharing of benefits derived from their use, which is especially relevant for microbial research involving the discovery and utilization of novel microorganisms (National Research Council et al. (NRC), 2004).

Another important international tool is the Cartagena Protocol on Biosafety, which addresses the safe transfer, handling, and use of living modified organisms (LMOs) resulting from modern biotechnology. This protocol requires countries to implement biosafety measures to mitigate potential risks to biodiversity and human health from LMOs, including genetically modified microorganisms (GMMs) (NRC, 2004). The protocol also includes the Advanced Informed Agreement (AIA) procedure, which requires exporting countries to notify and obtain consent from importing countries before shipping LMOs that may harm biodiversity.

The World Health Organization (WHO) offers additional guidance on laboratory biosafety and biosecurity, most notably through its Laboratory Biosafety Manual, which establishes global standards for the safe handling of pathogenic microbes. The WHO also coordinates international efforts to prevent the use of microbial research for bioterrorism or biological warfare, as outlined in the Biological Weapons Convention (BWC) (NRC, 2004).

10.3.2. National Legislation and Regulations

National governments play an important role in regulating microbial research by enacting legislation that upholds international agreements and addresses specific national concerns. In the United States, for example, microbial research is governed by a comprehensive set of laws and regulations enforced by federal agencies such as the National Institutes of Health (NIH), the Centers for Disease Control and Prevention (CDC), the Environmental Protection Agency (EPA), and the Food and Drug Administration (FDA) (Meechan and Potts, 2020; Desikan et al., 2011). The National Institutes of Health Guidelines for Research Involving Recombinant or Synthetic Nucleic Acid Molecules establish the safety standards for genetic engineering research, including the use of GMMs. These guidelines establish BSLs and specify containment practices for laboratories that handle various microorganism classes (NRC, 2004).

The CDC's Biosafety in Microbiological and Biomedical Laboratories (BMBL) manual provides detailed guidance on biosafety practices, such as the classification of infectious agents into risk groups, biosafety level implementation, and laboratory facility, equipment, and personnel training requirements (NRC, 2004). The EPA regulates the environmental release of GMOs to ensure that such activities do not endanger ecosystems or public health. The FDA regulates the use of microorganisms in the development of drugs, vaccines, and food products to ensure their safety and effectiveness.

In Europe, the European Union (EU) has created a strong regulatory framework for microbial research, particularly through directives and regulations governing the use of GMOs and pathogens. The EU Directive 2009/41/EC on the contained use of GMMs specifies the risk assessment, containment measures, and emergency response procedures for laboratories that work with GMMs. The directive also requires member states to establish competent authorities to monitor the implementation of these regulations and ensure compliance with EU standards (Desikan et al., 2011).

10.3.3. Institutional Oversight and Ethical Review

In addition to international and national regulations, microbial research is subject to institutional oversight and ethical review processes that ensure regulatory compliance and ethical standards. Universities, hospitals, and private laboratories typically form Institutional Biosafety Committees (IBCs) to oversee microbial research activities. These committees are in charge of reviewing research proposals, assessing risks, and ensuring that adequate biosafety measures are in place (NRC, 2004).

IBCs also play an important role in the ethical oversight of microbial research, ensuring that studies involving humans or animals follow ethical principles such as autonomy, beneficence, nonmaleficence, and justice. This includes obtaining research participants' informed consent, protecting their privacy and confidentiality, and minimizing harm to research animals (Woodworth et al., 2017).

In the United States, the NIH Guidelines require that institutions receiving NIH funding from IBCs to oversee research involving recombinant or synthetic nucleic acid molecules. These committees must include biosafety, microbiology, and related field experts, as well as community representatives, to address public concerns (Woodworth et al., 2017).

10.3.4. Ethical Guidelines and Codes of Conduct

Ethical considerations are central to the regulation of microbial research, guiding scientists' responsible behavior and ensuring that research is conducted in a way that respects human rights, animal welfare, and environmental integrity. Professional organizations, such as the American Society for Microbiology (ASM), frequently develop ethical guidelines and codes of conduct. ASM's code of ethics emphasizes the importance of integrity, transparency, and accountability in microbiological research (NRC, 2004).

These guidelines frequently cover topics such as the responsible use of genetic engineering technologies, the prevention of dual-use research of concern (DURC), and the ethical implications of synthetic biology. Researchers are encouraged to engage in ongoing ethical reflection, taking into account the broader societal implications of their work and taking proactive steps to reduce potential risks (Desikan et al., 2011).

10.4. INTELLECTUAL PROPERTY RIGHTS IN MICROBIOLOGY

Intellectual property rights (IPR) represent a significant ethical consideration in the regulation of microbiology research and practice. As advancements in microbiology increasingly intersect with biotechnology, genetics, and pharmaceuticals, the protection of intellectual property has become essential for fostering innovation, encouraging investment, and ensuring that the benefits of research are appropriately distributed (Desikan et al., 2011). However, the application of IPR in microbiology also raises complex ethical issues related to access, equity, and the balance between public and private interests.

10.4.1. Intellectual Property Rights

Patents are common examples of intellectual property rights, as they grant inventors or institutions exclusive rights to new inventions such as genetically engineered microorganisms, biotechnological processes, and novel therapeutic agents. Patents grant the patent holder the exclusive right to use, manufacture, sell, or license the invention for a set period, typically 20 years. This exclusivity is intended to encourage innovation by allowing inventors to recoup the costs of RandD and profit from their inventions (Singh et al., 2016). However, the use of patents in microbiology, particularly for living organisms and genetic materials, raises many ethical issues. The ability to patent life forms, such as GMMs or DNA sequences, has sparked heated debate. Critics argue that life should not be commodified or owned because doing so could limit access to critical technologies and knowledge, particularly in medicine and agriculture (Woodworth et al., 2017). Furthermore, the patenting of microorganisms and genetic resources can create barriers to research, as scientists may need to obtain licenses or pay royalties to use patented materials, potentially impeding future innovation.

10.4.2. Access and Equity

One of the most pressing ethical concerns about intellectual property rights in microbiology is access and equity. Patents can create monopolies that limit access to critical technologies, especially in low- and middle-income countries (LMICs), where the cost of patented products, such as life-saving drugs or diagnostic tools, can be prohibitively expensive. This may exacerbate health disparities and jeopardize global public health efforts. For example, patenting antiretroviral drugs used to treat HIV/AIDS initially restricted access to these life-saving medications in many developing countries, sparking international debates about the ethical implications of intellectual property rights in the context of global health (Singh et al., 2016).

Ethical frameworks, such as the Doha Declaration on the TRIPS Agreement and Public Health (2001), emphasize the importance of balancing intellectual property rights with the right to access essential medicines. The WTO adopted this declaration, which affirms the flexibility of TRIPS (Trade-Related Aspects of Intellectual Property Rights) provisions by allowing member states to prioritize public health and access to medicines. It supports the use of compulsory licensing and parallel imports to improve access to affordable drugs during public health emergencies (NRC, 2004).

10.4.3. Benefit Sharing and Biodiversity

The ethical consideration of benefit sharing is especially important in microbial research that uses genetic resources derived from biodiversity hotspots, which are frequently located in developing countries. The Nagoya Protocol on Access and Benefit Sharing (ABS) of the Convention on Biological Diversity (CBD) addresses the ethical and legal implications of accessing genetic resources and sharing the benefits that result from their use. The protocol seeks to ensure that countries and communities that provide genetic resources receive a fair share of the benefits, such as financial compensation, technology transfer, and capacity building (Singh et al., 2016).

However, implementing the Nagoya Protocol has proven difficult, particularly in balancing patent holders' interests with the rights of indigenous and local communities. Ethical concerns arise when bioprospecting, or the exploration of biodiversity for commercially valuable genetic and biochemical resources, results in the exploitation of these communities without adequate compensation or recognition (Meechan and Potts, 2020). Ensuring equitable benefit sharing necessitates strong legal frameworks and active participation of local communities in decision-making processes involving the use of their genetic resources.

10.4.4. Balancing Public and Private Interests

The ethical regulation of IPR in microbiology requires a careful balance of public and private interests. On the one hand, patents are critical for encouraging investment in R&D, especially in industries where innovation costs are high, such as pharmaceuticals and biotechnology. On the other hand, the public interest requires that critical technologies, particularly those related to health and food security, remain accessible and affordable (Singh et al., 2016).

Ethical approaches to IPR in microbiology emphasize the importance of transparency, fairness, and public accountability in the patenting process. This includes ensuring that patents are only granted for truly novel and non-obvious inventions and that patent holders do not use their rights to create unfair monopolies. Furthermore, ethical considerations necessitate mechanisms that enable the sharing of patented technologies for humanitarian purposes, such as patent pools or open licensing agreements, which facilitate access to critical technologies in resource-constrained settings (Meechan and Potts, 2020).

10.5. BIOETHICS IN BIOTECHNOLOGY

As microbiology and biotechnology become more intertwined, particularly in areas such as genetic engineering, synthetic biology, and the development of GMOs, bioethics plays an increasingly important role in guiding these technologies' responsible applications (O'Mathúna, 2007). Bioethics ensures that microbiological biotechnological innovations respect human rights, environmental sustainability, and societal values.

10.5.1. Ethical Principles in Biotechnology

The application of bioethics in microbiology is based on four fundamental ethical principles: autonomy, beneficence, nonmaleficence, and justice. These principles are critical in guiding the regulation of biotechnological practices that involve microorganisms.

- **Autonomy**: The principle of autonomy emphasizes the importance of informed consent in microbiology research, especially when human subjects or communities are involved. For example, when conducting microbiome studies or developing microbial-based therapies, researchers must ensure that participants are fully informed about the nature, risks, and potential benefits of the study. This respects the autonomy of individuals and communities by allowing them to make informed decisions about their participation (Stasi and Thongpravati, 2024).
- **Beneficence and Non-maleficence**: Beneficence entails improving the well-being of individuals and society, whereas nonmaleficence entails avoiding harm. In microbiological biotechnology, these principles are especially important in the development and application of GMOs and synthetic microorganisms. The potential benefits of these innovations, such as enhanced agricultural productivity, new therapies, and environmental remediation, must be carefully weighed against the risks, including unintended ecological consequences, antibiotic resistance spread, and potential harm to human health (O'Mathúna, 2007).
- **Justice**: The principle of justice in microbiology is concerned with the equitable distribution of the benefits and burdens of biotechnological advancements. This includes ensuring that marginalized populations and developing countries have equal access to the benefits of microbial biotechnology, such as new infectious disease treatments or improved agricultural technology. Ethical regulation requires that the benefits of microbial research and biotechnology be shared widely to promote global health and well-being (Desikan et al., 2011).

10.5.2. Regulatory Frameworks and Bioethics

Bioethical concerns are deeply ingrained in the regulatory frameworks that govern microbiological biotechnology. International guidelines, such as UNESCO's Universal Declaration on Bioethics and Human Rights, set ethical standards for biotechnology research by emphasizing human dignity, human rights, and environmental sustainability (NRC, 2004). These guidelines are especially important for regulating microbial biotechnology, which has the potential to have a significant global impact.

National regulatory agencies, including the United States Food and Drug Administration (FDA) and the European Medicines Agency (EMA), incorporate bioethical principles into their oversight of microbiological biotechnology. This includes rigorous ethical review processes to determine the safety, efficacy, and potential societal impact of new microbial technologies. For example, the release of GMMs into the environment is subject to stringent regulatory oversight to prevent environmental damage and ensure public safety (Stasi and Thongpravati, 2024).

10.5.3. Challenges and Future Directions

Despite the robust role of bioethics in regulating microbiological biotechnology, challenges persist. Emerging technologies, such as CRISPR-Cas9 gene editing, raise new ethical questions about the extent of human intervention in microbial genomes and the potential long-term consequences of such interventions. Bioethics must evolve to address these challenges, ensuring that the regulation of microbiological biotechnology keeps pace with scientific advancements and aligns with ethical norms and societal values (O'Mathúna, 2007; Stasi and Thongpravati, 2024).

10.6. MICROBIAL PATENTING

Microbial patenting refers to the legal protection granted to innovations involving microorganisms, such as bacteria, fungi, and viruses, through patents. This practice plays a pivotal role in microbiology by safeguarding intellectual property and incentivizing research and development. Microbial patents cover various innovations, including GMMs, novel microbial strains, and processes involving microbes, contributing to advancements in biotechnology, pharmaceuticals, and environmental science.

10.6.1. Criteria for Patentability

To be patentable, a microbiological invention must meet certain broad characteristics, such as novelty, non-obviousness, and industrial usefulness.

Novelty requires that the microorganism or microbial process be new and not previously revealed in the public domain. Non-obviousness requires that the invention include an inventive step that is not obvious to someone with knowledge in the relevant field. Industrial application requires that the microbial invention be able to be used in a certain industry, such as agriculture, pharmaceuticals, or environmental management. Furthermore, if the patent contains a naturally existing microbe, it must have been isolated or modified in a way that adds value beyond its native state (De Mandal et al., 2022).

10.6.2. Types of Microbial Patents

Microbial patents are divided into numerous classes, each protecting a distinct component of the microbiological invention. The primary types are patents on microbial strains, genetically engineered microorganisms, microbial processes, and microbial products.

- **Microbial Strain Patents**: These patents protect naturally occurring or intentionally selected microbial strains that exhibit specified properties, such as increased antibiotic production or environmental resilience.
- **Genetically Created Microorganisms:** This category includes patents for microorganisms that have been genetically changed to show desirable features, such as bacteria created for bioremediation or yeast strains customized for biofuel production.
- **Microbial Processes**: These patents protect new microorganism-based processes, such as ways for manufacturing antibiotics, biofertilizers, and other biologically active substances.
- **Microbial Products**: Patents in this category protect the products derived from microbial activity, such as enzymes, metabolites, or biopolymers, which are utilized in various industrial applications (De Mandal et al., 2022).

10.6.3. Patent Application Process

The first step in the process of applying for a patent for a microbiological invention is to prepare and submit a comprehensive patent application. The microorganism, its traits, and the procedure or product it is connected to must all be clearly described in the application. Claims that specify the range of protection sought are frequently made in conjunction with this (Hewitt, 2004). A deposit of the microorganism in an authorized culture collection might be necessary if the innovation comprises a novel microbial strain to satisfy the legal criteria for reproducibility.

The patent office thoroughly examines the invention following submission to make sure it satisfies the requirements for patentability. In order to evaluate novelty and non-obviousness, this investigation looks at previous works of art. It may also entail technical analysis to confirm the microbiological invention's practical applicability. The patent office may ask for more information or changes to the claims during the course of the examination process, which could involve several rounds. The patent is awarded to the creator, giving them exclusive rights to the microbiological invention for a predetermined amount of time, usually 20 years from the filing date, after the investigation is over and the invention is found to be patentable (Hewitt, 2004).

10.6.4. Examples of Microbial Patents

The variety and significance of microbial inventions are demonstrated by a number of notable instances of microbial patents. The patent on the soil bacteria *Streptomyces avermitilis*, which is used to make the antiparasitic medication ivermectin, is one well-known example. The treatment of parasite illnesses in both humans and animals has been greatly impacted by this patent.

An additional illustration would be the patent for the genetically altered bacterium *Pseudomonas putida*, which was the first organism to be genetically changed and granted a patent in the US. This bacteria is useful for environmental bioremediation of oil spills since it was created to degrade hydrocarbons. The creation of Bt crops that are resistant to specific pests has been facilitated by the patent for Bacillus thuringiensis (Bt) strains, which generate insecticidal proteins. This has had a major positive impact on crop protection and production in agriculture.

These illustrations show how diverse the uses of microbial patents are in the fields of agriculture, medicine, and environmental management, underscoring the need for legal protection in promoting microbiological innovation (Lederberg and Relman, 2009).

10.7. ETHICAL ISSUES IN GENETIC ENGINEERING

Genetic engineering in microbiology has transformed the discipline by allowing for precise alterations to microbial genomes, resulting in substantial advances in medicine, agriculture, and industry. However, these great powers present difficult ethical concerns that must be carefully addressed. Concerns regarding potential hazards to human health, environmental sustainability, and social equality are central to the ethical discussion surrounding genetic engineering.

Addressing these ethical quandaries is critical to ensuring the responsible and equitable application of genetic technologies.

10.7.1. Ethical Considerations in Genetic Engineering

The ethical considerations of genetic engineering are primarily concerned with the potential repercussions of changing microbial genomes. One of the most fundamental ethical concerns is whether humans have the authority to change the genetic makeup of species, particularly when the long-term consequences are unknown. The prospect of developing GMOs capable of outcompeting natural species raises concerns about playing 'God' with nature. Furthermore, the possibility of unforeseen outcomes, such as the evolution of new pathogens or the transmission of antibiotic-resistance genes, needs a cautious approach. The ethical precept "do no harm" should govern genetic engineering research and uses, ensuring that potential benefits exceed hazards (Woodworth et al., 2017).

10.7.2. Risk Assessment and Management

Risk assessment and management are essential elements of ethical genetic engineering. Unintended ecological consequences of genetic alterations in microorganisms include the disruption of native microbial communities and the transmission of synthetic features to wild populations. To reduce these dangers, comprehensive risk assessment processes must be developed that incorporate both scientific evidence and ethical considerations. This process entails assessing the possibility of unfavorable consequences and devising methods to avoid or mitigate harm. Furthermore, risk management should be an ongoing process, with genetically altered microbes being monitored and evaluated in real-world contexts to verify their safety and efficacy (Desikan et al., 2011).

10.7.3. Regulatory and Legal Framework

A strong regulatory and legal framework is required to ensure the ethical use of genetic engineering in microbiology. International and national rules must be enacted to govern the development, testing, and deployment of genetically altered microbes. These rules should be based on scientific facts and ethical values, ensuring that genetic changes are safe for people and the environment. Furthermore, legal frameworks should address intellectual property issues, such as patenting GMOs, while also ensuring that these technologies are accessible and cheap to everybody. Clear and transparent criteria will help to foster public trust and responsible innovation in genetic engineering (Woodworth et al., 2017).

10.7.4. Implications for Biodiversity and Ecosystem

Ecosystems and biodiversity are significantly impacted by genetic engineering. Unintended consequences of introducing genetically engineered microbes into the environment include alterations to ecosystem dynamics and a decrease in genetic diversity. For instance, native species may be displaced by artificial bacteria created to break down contaminants, which would result in a decline in biodiversity. Furthermore, there's a chance that modified features will expand beyond their intended objectives due to horizontal gene transfer (HGT) between modified and normal animals. Before releasing GMOs into the environment, it is crucial to carry out comprehensive ecological evaluations to maintain biodiversity. Additionally, systems for monitoring and minimizing any negative effects must be developed (Desikan et al., 2011).

10.7.5. Equity and Access to Genetic Technologies

Equity and access to genetic tools are important ethical concerns in genetic engineering. The advantages of genetic engineering should be dispersed evenly among diverse populations and regions, preventing inequities between those who have access to these technologies and those who do not. Developing countries and underprivileged people frequently lack the resources to benefit from advances in genetic engineering, potentially leading to disparities in health, agriculture, and economic possibilities. As a result, laws should be created to ensure that genetic technologies are available to everyone, regardless of socioeconomic status. This involves fostering technological transfer, making genetic engineering techniques cheaper, and incorporating a wide range of stakeholders in decision-making (Desikan et al., 2011).

10.8. MICROBIAL DATA SHARING AND PRIVACY

Microbial data sharing and privacy are critical concerns in the field of microbiology, as they influence the advancement of research, public health outcomes, and ethical considerations. The proliferation of high-throughput sequencing technologies and global collaboration in research has exponentially increased the volume of microbial data generated. This necessitates a robust framework to address the challenges of data privacy, confidentiality, and secure sharing practices.

10.8.1. Types of Microbial Data

- Microbial data can take numerous forms, including genomic, metagenomic, and phenotypic data. Genomic data consists of entire DNA sequences of microorganisms, which provide insights into their genetic makeup and potential capabilities.

- Metagenomic data is generated by sequencing the collective genomes of microbial communities in environmental or clinical samples, allowing for a more comprehensive understanding of microbial diversity and interactions.
- Phenotypic data refers to microorganisms' observable features or qualities, such as growth patterns, resistance profiles, and metabolic capacities. Each sort of data is important in microbiological research, but it presents unique issues in terms of privacy and sharing (Woodworth et al., 2017).

10.8.2. Data Privacy and Confidentiality

In microbiological research, data confidentiality and privacy are crucial, especially when handling sensitive data. When microbiological data is connected to private health records, privacy issues occur since it may reveal identifiers about specific people or groups. Confidentiality entails protecting information from unwanted access and making sure that private data is shared only when necessary (Gordon et al., 2013). To reduce privacy issues, methods like data anonymization and de-identification are used. However, keeping information private while allowing for data sharing is a difficult task, particularly given the growing sophistication of data re-identification methods (Gordon et al., 2013).

10.8.3. Ethical and Legal Framework

The legal and ethical framework that controls the sharing of microbiological data is intended to strike a compromise between the need to preserve privacy and the advantages of open access. Ethically, researchers need to make sure that participants or contributors give their informed consent, outlining the uses, storage, and sharing of their data. Legally speaking, requirements for data protection and privacy are established by many laws, including the Health Insurance Portability and Accountability Act (HIPAA) in the US and the General Data Protection Regulation (GDPR) in the EU (Gordon et al., 2013). Maintaining public trust and avoiding legal ramifications depend on compliance with these standards.

10.8.4. Data Sharing Policies and Best Practices

Promoting collaboration while protecting data requires the implementation of effective data-sharing policies and best practices. Policies should include explicit guidelines for data access, usage, and sharing to guarantee that data is made available in a way that respects privacy and meets legal obligations. Best practices include employing data-sharing agreements that outline terms of use, storing data in reliable repositories with restricted access, and putting data management strategies that take privacy issues into account. Therefore, promoting the use of standardized formats and metadata can help with data sharing and cross-platform interoperability (Meechan and Potts, 2020).

10.8.5. Data Security and Management

Data security and management are critical for protecting microbiological data against unauthorized access and intrusions. Data security entails using technical safeguards like encryption, secure access controls, and frequent audits to protect data. Data management includes techniques including regular backups, version control, and extensive documentation to assure data integrity and dependability. Establishing strong processes for data protection and management is critical to retaining data contributors' trust and the integrity of research results (Meechan and Potts, 2020).

10.9. MICROBIAL RESEARCH MISCONDUCT

Microbial research in microbiology plays a crucial role in understanding pathogens, developing treatments, and advancing biotechnology. However, research misconduct within this field poses significant risks to scientific integrity and public health. Misconduct in microbial research, such as data fabrication, falsification, and plagiarism, undermines the credibility of scientific findings and can have far-reaching consequences. Addressing these issues is vital to ensuring the reliability of scientific research and maintaining public trust in microbiology.

10.9.1. Commonest Types of Misconduct

The most common forms of research misconduct in microbiology include data fabrication, falsification, and plagiarism.

- Data fabrication is the process of creating non-existent data or outcomes and then presenting them as genuine findings. This is especially detrimental in microbial research, as false data may lead to inaccurate conclusions concerning the behavior or treatment of microorganisms.
- Data falsification is the alteration of research data and methods, frequently to ensure that the results match the desired outcome. This can include modifying experimental circumstances, removing specific data points, or adjusting the statistical analysis to support a particular hypothesis.
- Plagiarism is another common sort of misbehavior in which researchers use someone else's work or ideas without giving proper

credit. In microbiology, this could include unlawful use of another scientist's research methodology, findings, or even entire sections of a manuscript. Each of these misconducts not only corrupts the research process but also undermines the scientific community's integrity (Stasi and Thongpravati, 2024).

10.9.2. Preventive Measures and Best Practices

In order to prevent research misconduct in microbiology, strong standards and best practices must be implemented. One of the most effective ways is to create explicit ethical norms for all researchers. Research organizations should enforce these principles, which explain appropriate practices, data management protocols, and the need for accurate attribution. Education and training programs on research ethics should be required for all researchers, especially those in the early phases of their careers. Such activities can help inculcate a firm grasp of ethical research techniques as well as the value of scientific integrity.

Another protective approach is to implement effective peer-review processes. Peer review serves as a key checkpoint, ensuring that research findings are thoroughly reviewed by experts in the field before publishing. Data management practices play an important role in preventing wrongdoing. Researchers should be expected to keep detailed and transparent records of their experimental protocols, data collection methods, and analytical approaches. This can be helped by using electronic lab notebooks and data repositories, which provide an auditable record of research activity. Furthermore, creating an open and collaborative research environment can lessen the pressure to engage in wrongdoing by encouraging shared ownership and mutual accountability among researchers (Baltz et al., 2010; Desikan et al., 2011).

10.9.3. Detection and Reporting Mechanism

Maintaining scientific integrity in microbiology requires the detection and reporting of research misconduct. There are several ways to find plagiarism, such as data audits, peer review, and plagiarism detection software. Peer reviewers are frequently in the greatest position to spot irregularities or questionable data in research publications because they are subject matter experts in the field. By contrasting the stated results with the raw data, data audits, carried out by impartial organizations, can assist in confirming the validity of the claimed results. Research papers are increasingly being examined for plagiarized content using tools like *Turnitin* and *iThenticate*.

Research institutions should provide accessible and confidential reporting channels for the reporting of misconduct. Protecting whistleblowers from reprisal is essential to promoting the reporting of unethical activity. Ethics committees and institutional review boards (IRBs) ought to have the authority to look into accusations of wrongdoing in-depth and objectively, making sure that due process is adhered to in every situation (Baltz et al., 2010).

10.9.4. Consequences and Penalties

Research misconduct in microbiology can have serious repercussions for the participants as well as the larger scientific community. If found guilty of wrongdoing, researchers may have professional repercussions such as having their published articles retracted, losing their funding for research, and having their reputations damaged. Under certain circumstances, researchers may also be subject to legal repercussions, especially if their malfeasance has caused financial losses or harm to the public. Disciplinary actions such as job termination, suspension, or banishment from further research endeavors may be enforced by institutions. Misconduct can damage public confidence in science and impede the advancement of significant scientific breakthroughs, which has a negative impact on the scientific community as a whole (Meechan and Potts, 2020).

10.9.5. Promoting a Culture of Integrity

The scientific community, funding organizations, and research institutions must continue to support an environment of honesty in microbial research. Leadership in organizations ought to stress the value of moral research procedures and set an example of this conduct in their own work. Integrity can also be enhanced by transparency in research procedures, such as providing open access to data and methodologies since it enables independent validation of study conclusions. Recognition and reward systems can encourage adherence to high ethical standards, and mentoring programs can help early-career researchers implant ethical ideals. It is possible to reduce the likelihood of misbehavior and guarantee the continuous progress of microbiology in the scientific community by cultivating an atmosphere that values and rewards integrity (Baltz et al., 2010).

10.10. MICROBIAL ETHICS IN THE PUBLIC SPHERE

Microbial ethics in the public sphere encompasses the complex interplay between scientific advancements in microbiology and societal values. As microbiological research and applications expand, particularly in biotechnology and environmental sciences, ethical considerations become crucial. These include maintaining public trust, navigating regulatory frameworks, and ensuring equitable access to innovations. Addressing these issues is vital for fostering responsible science and protecting public interests.

10.10.1. Public Perception and Trust

Public perception and trust have a tremendous impact on the acceptability and success of microbial research and biotechnology technologies. Media coverage, educational outreach, and previous conflicts can all influence how the public perceives scientific procedures such as genetic alterations or microbial bioremediation. Transparency and clear communication regarding the benefits, hazards, and uncertainties connected with microbial technologies are critical for developing and retaining confidence. Engaging with the public through forums, consultations, and educational programs helps to clarify scientific concepts and answer concerns, increasing the credibility and acceptance of microbial research (NRC, 2004).

10.10.2. Regulatory and Policy Frameworks

Effective regulatory and policy frameworks are critical for managing microbial research and applications, ensuring that scientific progress is made responsibly. In the United States, institutions like as the EPA and the (FDA monitor the safety and efficacy of microbial technologies. International frameworks, such as the Cartagena Protocol on Biosafety, govern the transport and use of GMOs across borders (UNEP, 2000). Policies must strike a balance between innovation and risk management, addressing safety concerns while not limiting scientific development. Regular revisions to these frameworks are required to keep up with the rapid advances in microbial science and technology (National Research Council et al. (NRC), 2004).

10.10.3. Ethical Issues in Biotechnology Applications

Significant concerns are raised by ethical difficulties in biotechnology applications, especially when it comes to GMMs and synthetic biology. For example, discussions concerning the possible environmental repercussions of GMM use in agriculture, such as unintentional effects on non-target species and ecological balance, are sparked. Comparably, concerns concerning the origins of life forms and their effects on natural ecosystems are raised by synthetic biology, which entails the design and construction of novel biological components. To ensure that biotechnological applications are developed and deployed in a socially acceptable manner, addressing these ethical concerns calls for thorough risk assessments, ethical discussions, and stakeholder engagement (Gibson et al., 2010).

10.10.4. Equity and Access to Microbial Innovations

Equity and access to microbiological breakthroughs are crucial issues in the public domain, as discrepancies might result in unequal gains from scientific advances. Ensuring fair access to microbial technologies like vaccinations and therapeutic treatments is critical for global health (Cohen, 2010). This includes resolving hurdles to economic resources, infrastructure, and intellectual property rights that can impede access in low- and middle-income nations. Policies that promote equitable distribution and coordinated efforts between developed and developing countries can assist in reducing these discrepancies. Fostering open-access programs and equitable license arrangements can also help to disseminate the advantages of microbial breakthroughs (Morse, 2012).

10.10.5. Informed Consent and Community Engagement

Informed consent and community engagement are critical components of ethical microbiological research. Researchers must seek participants' informed consent by fully explaining the nature of the study, potential hazards, and the purpose of any gathered data (World Health Organization, 2011). Community engagement is actively engaging stakeholders in talks regarding study objectives, prospective outcomes, and ethical considerations. This participatory strategy ensures that research reflects community values and addresses local needs. Effective community participation enhances transparency and increases public trust in scientific research and its applications (Morse, 2012).

BIBLIOGRAPHY

1. Abbasi, B., Harper, J., & Ahmadvand, S. (2021). A short critique on biomining technology for critical materials. *World Journal of Microbiology and Biotechnology, 37*(5). https://doi.org/10.1007/s11274-021-03048-9.
2. Allison, S. D., Wallenstein, M. D., & Bradford, M. A. (2010). Soil-carbon response to warming dependent on microbial physiology. *Nature Geoscience, 3*(5), 336–340. https://doi.org/10.1038/ngeo846.
3. Alonso, S. (2016). Novel preservation techniques for microbial cultures. In K. S. Ojha & B. K. Tiwari (Eds.), *Novel Food Fermentation Technologies* (pp. 7–33). Springer International Publishing. https://doi.org/10.1007/978-3-319-42457-6_2.
4. Amin, S. A., Hmelo, L. R., van Tol, H. M., Durham, B. P., Carlson, L. T., Heal, K. R., Morales, R. L., Berthiaume, C. T., Parker, M. S., Djunaedi, B., Ingalls, A. E., Parsek, M. R., Moran, M. A., & Armbrust, E. V. (2015). Interaction and signalling between a cosmopolitan phytoplankton and associated bacteria. *Nature, 522*(7554), 98–101. https://doi.org/10.1038/nature14488.
5. Anderson, T. H. (2003). Microbial eco-physiological indicators to assess soil quality. *Agriculture, Ecosystems & Environment, 98*(1), 285–293. https://doi.org/10.1016/S0167-8809(03)00088-4.
6. Andler, R., Tiso, T., Blank, L., Andreeßen, C., Zampolli, J., D'Afonseca, V., Guajardo, C., & Díaz-Barrera, A. (2022). Current progress on the biodegradation of synthetic plastics: From fundamentals to biotechnological applications. *Reviews in Environmental Science and Bio/Technology, 21*(4), 829–850. https://doi.org/10.1007/s11157-022-09631-2.

7. Anno, F. D., Rastelli, E., Sansone, C., Brunet, C., Ianora, A., & Anno, A. D. (2021). Bacteria, fungi, and microalgae for the bioremediation of marine sediments contaminated by petroleum hydrocarbons in the omics era. *Microorganisms, 9*(8), 1695. https://doi.org/10.3390/microorganisms9081695.
8. Atlas, R. M. (2010). *Handbook of Microbiological Media* (4th ed.). Taylor & Francis.
9. Baltz, R. H., Demain, A. L., & Davies, J. E. (2010). *Manual of Industrial Microbiology and Biotechnology*. American Society for Microbiology Press.
10. Beveridge, T. J., Lawrence, J. R., & Murray, R. G. E. (2007). Sampling and staining for light microscopy. In *Methods for General and Molecular Microbiology* (pp. 19–33). John Wiley & Sons, Ltd. https://doi.org/10.1128/9781555817497.ch2.
11. Bolon, B., Uzal, F. A., & Schutten, M. (2023). Bacterial toxins. In *Elsevier eBooks* (pp. 629–678). https://doi.org/10.1016/b978-0-443-16153-7.00009-5.
12. Bonkowski, M. (2004). Protozoa and plant growth: The microbial loop in soil revisited. *New Phytologist, 162*(3), 617–631. https://doi.org/10.1111/j.1469-8137.2004.01066.x.
13. Borges, F., Briandet, R., Callon, C., Champomier-Vergès, M. C., Christieans, S., Chuzeville, S., Denis, C., Desmasures, N., Desmonts, M. H., Feurer, C., Leroi, F., Leroy, S., Mounier, J., Passerini, D., Pilet, M. F., Schlusselhuber, M., Stahl, V., Strub, C., Talon, R., & Zagorec, M. (2022). Contribution of omics to biopreservation: Toward food microbiome engineering. *Frontiers in Microbiology, 13*. https://doi.org/10.3389/fmicb.2022.951182.
14. Boto, L. (2010). Horizontal gene transfer in evolution: Facts and challenges. *Proceedings of the Royal Society B: Biological Sciences, 277*(1683), 819–827. https://doi.org/10.1098/rspb.2009.1679.
15. BYJUS. (2021). *Bioremediation–Process | Types of Bioremediation | Examples*. Retrieved from: https://byjus.com/biology/bioremediation/ (accessed on 9 November 2024).
16. BYJUS. (2022a). *Microbiology–All About Microorganisms and Its Different Branches*. Retrieved from: https://byjus.com/biology/microbiology/ (accessed on 9 November 2024).
17. BYJUS. (2022b). *Kingdom Fungi- Structure, Characteristics & Classification of Fungi*. Retrieved from: https://byjus.com/biology/kingdom-fungi/ (accessed on 9 November 2024).

18. Callieri, C., Eckert, E. M., Di Cesare, A., & Bertoni, F. (2019). Microbial communities. In B. Fath (Ed.), *Encyclopedia of Ecology* (2nd ed., pp. 126–134). Elsevier. https://doi.org/10.1016/B978-0-12-409548-9.11222-9.

19. Chatterjee, S., & Raval, I. H. (2019). Pathogenic microbial genetic diversity with reference to health. In S. Das & H. R. Dash (Eds.), *Microbial Diversity in the Genomic Era* (pp. 559–577). Academic Press. https://doi.org/10.1016/B978-0-12-814849-5.00032-0.

20. Chaudhary, D. K., Khulan, A., & Kim, J. (2019). Development of a novel cultivation technique for uncultured soil bacteria. *Scientific Reports, 9*(1). https://doi.org/10.1038/s41598-019-43182-x.

21. Cheung, G. Y. C., Bae, J. S., & Otto, M. (2021). Pathogenicity and virulence of *Staphylococcus aureus*. *Virulence, 12*(1), 547–569. https://doi.org/10.1080/21505594.2021.1878688.

22. Conrad, R. (2009). The global methane cycle: Recent advances in understanding the microbial processes involved. *Environmental Microbiology Reports, 1*(5), 285–292. https://doi.org/10.1111/j.1758-2229.2009.00038.x.

23. Cordell, G. A. (2000). Biodiversity and drug discovery—A symbiotic relationship. *Phytochemistry, 55*(6), 463–480. https://doi.org/10.1016/S0031-9422(00)00230-2.

24. Da Silva, N., Taniwaki, M. H., Junqueira, V. C. A., De Arruda Silveira, N. F., Okazaki, M. M., & Gomes, R. A. R. (2018). *Microbiological Examination Methods of Food and Water: A Laboratory Manual* (2nd ed.). CRC Press. https://doi.org/10.1201/9781315165011.

25. De Mandal, S., Panda, A. K., Senthil Kumar, N., Bisht, S. S., & Jin, F. (2022). *Metagenomics and Microbial Ecology: Techniques and Applications* (1st ed.). CRC Press.

26. Denecke, M. A., Bryan, N., Kalmykov, S., Morris, K., & Quinto, F. (2018). Sources and behaviour of actinide elements in the environment. In J. K. Gibson & W. A. Jong (Eds.), *Experimental and Theoretical Approaches to Actinide Chemistry* (pp. 378–444). Wiley. https://doi.org/10.1002/9781119115557.ch8.

27. Dennehy, J. J. (2016). Evolutionary ecology of virus emergence. *Annals of the New York Academy of Sciences, 1389*(1), 124–146. https://doi.org/10.1111/nyas.13304.

28. Desikan, P., Chakrabarti, A., & Muthuswamy, V. (2011). Ethical issues in microbiology. *Indian Journal of Medical Microbiology, 29*(4), 327–330. https://doi.org/10.4103/0255-0857.90154.

29. Du, Q., Zhang, D., Hu, W., Li, X., Xia, Q., Wen, T., & Jia, H. (2021). Nosocomial infection of COVID-19: A new challenge for healthcare professionals (Review). *International Journal of Molecular Medicine, 47*(4). https://doi.org/10.3892/ijmm.2021.4864.

30. Enwall, K., Nyberg, K., Bertilsson, S., Cederlund, H., Stenström, J., & Hallin, S. (2007). Long-term impact of fertilization on activity and composition of bacterial communities and metabolic guilds in agricultural soil. *Soil Biology and Biochemistry, 39*(1), 106–115. https://doi.org/10.1016/j.soilbio.2006.06.015.

31. Epstein, S. (2013). The phenomenon of microbial uncultivability. *Current Opinion in Microbiology, 16*(5), 636–642. https://doi.org/10.1016/j.mib.2013.08.003.

32. Fierer, N., Nemergut, D., Knight, R., & Craine, J. M. (2010). Changes through time: Integrating microorganisms into the study of succession. *Research in Microbiology, 161*(8), 635–642. https://doi.org/10.1016/j.resmic.2010.06.002.

33. Fortier, L. C., & Sekulovic, O. (2013). Importance of prophages to evolution and virulence of bacterial pathogens. *Virulence, 4*(5), 354–365. https://doi.org/10.4161/viru.24498.

34. Gautam, S. S., Gondwal, M., Soni, R., & Gautam, B. P. S. (2022). Potash biofertilizers: Current development, formulation, and applications. In R. Soni, D. C. Suyal, A. N. Yadav, & R. Goel (Eds.), *Trends of Applied Microbiology for Sustainable Economy* (pp. 481–500). Academic Press. https://doi.org/10.1016/B978-0-323-91595-3.00014-8.

35. GeeksforGeeks. (2023). *Virus in Biology*. GeeksforGeeks. https://www.geeksforgeeks.org/what-is-virus/ (accessed on 9 November 2024).

36. GeeksforGeeks. (2024a). *Bacterial Genetics*. GeeksforGeeks. https://www.geeksforgeeks.org/bacterial-genetics/ (accessed on 9 November 2024).

37. GeeksforGeeks. (2024b). *Facts About Algae*. GeeksforGeeks. https://www.geeksforgeeks.org/facts-about-algae/ (accessed on 9 November 2024).

38. Gest, H. (2004). The discovery of microorganisms by Robert Hooke and Antoni Van Leeuwenhoek, fellows of the Royal Society. *Notes and Records of the Royal Society of London, 58*(2), 187–201. https://doi.org/10.1098/rsnr.2004.0055.

39. Gilmore, B. F., & Denyer, S. P. (2023). *Hugo and Russell's Pharmaceutical Microbiology*. John Wiley & Sons.

40. Głodowska, M., & Wozniak, M. (2019). Changes in soil microbial activity

and community composition as a result of selected agricultural practices. *Agricultural Sciences, 10*(03), 330–351. https://doi.org/10.4236/as.2019.103028.

41. Gougoulias, C., Clark, J. M., & Shaw, L. J. (2014). The role of soil microbes in the global carbon cycle: Tracking the below-ground microbial processing of plant-derived carbon for manipulating carbon dynamics in agricultural systems. *Journal of the Science of Food and Agriculture, 94*(12), 2362–2371. https://doi.org/10.1002/jsfa.6577.
42. Hall, R. (2015). Protozoa. *AccessScience*. https://doi.org/10.1036/1097-8542.551900.
43. Hewitt, W. (2004). *Microbiological Assay for Pharmaceutical Analysis: A Rational Approach*. Interpharm/CRC.
44. Holmes, A. H., Moore, L. S. P., Sundsfjord, A., Steinbakk, M., Regmi, S., Karkey, A., Guerin, P. J., & Piddock, L. J. V. (2016). Understanding the mechanisms and drivers of antimicrobial resistance. *The Lancet, 387*(10014), 176–187. https://doi.org/10.1016/S0140-6736(15)00473-0.
45. Hurst, C. J., Crawford, R. L., Garland, J. L., & Lipson, D. A. (2007). *Manual of Environmental Microbiology*. American Society for Microbiology Press.
46. Hyde, K., Xu, J., Rapior, S., Jeewon, R., Lumyong, S., Niego, A. G. T., Abeywickrama, P., Aluthmuhandiram, J., Brahamanage, R., Brooks, S., Chaiyasen, A., Chethana, K., Chomnunti, P., Chepkirui, C., Chuankid, B., De Silva, N., Doilom, M., Faulds, C., Gentekaki, E., & Stadler, M. (2019). The amazing potential of fungi: 50 ways we can exploit fungi industrially. *Fungal Diversity, 97*(1), 1–136. https://doi.org/10.1007/s13225-019-00430-9.
47. Jagadeesan, B., Gerner-Smidt, P., Allard, M. W., Leuillet, S., Winkler, A., Xiao, Y., Chaffron, S., Van Der Vossen, J., Tang, S., Katase, M., McClure, P., Kimura, B., Ching Chai, L., Chapman, J., & Grant, K. (2019). The use of next generation sequencing for improving food safety: Translation into practice. *Food Microbiology, 79*, 96–115. https://doi.org/10.1016/j.fm.2018.11.005.
48. Jaiswal, S., & Shukla, P. (2020). Alternative strategies for microbial remediation of pollutants via synthetic biology. *Frontiers in Microbiology, 11*. https://doi.org/10.3389/fmicb.2020.00808.
49. Janssen, R., Krogfelt, K. A., Cawthraw, S. A., Van Pelt, W., Wagenaar, J. A., & Owen, R. J. (2008). Host-pathogen interactions in *Campylobacter* infections: The host perspective. *Clinical Microbiology Reviews, 21*(3), 505–518. https://doi.org/10.1128/cmr.00055-07.

50. Jasson, V., Jacxsens, L., Luning, P., Rajkovic, A., & Uyttendaele, M. (2010). Alternative microbial methods: An overview and selection criteria. *Food Microbiology, 27*(6), 710–730. https://doi.org/10.1016/j.fm.2010.04.008.

51. Jinek, M., Chylinski, K., Fonfara, I., Hauer, M., Doudna, J. A., & Charpentier, E. (2012). A programmable dual-RNA–guided DNA endonuclease in adaptive bacterial immunity. *Science, 337*(6096), 816–821. https://doi.org/10.1126/science.1225829.

52. Johnson, D. B. (2014). Biomining—Biotechnologies for extracting and recovering metals from ores and waste materials. *Current Opinion in Biotechnology, 30*, 24–31. https://doi.org/10.1016/j.copbio.2014.04.008.

53. Journal of Biodiversity and Environmental Sciences | JBES. (2024). *International Network for Natural Sciences | Research Journal*. https://innspub.net/journal-of-biodiversity-and-environmental-sciences/ (accessed on 9 November 2024).

54. Jüttner, M., & Ferreira-Cerca, S. (2022). Looking through the lens of the ribosome biogenesis evolutionary history: Possible implications for archaeal phylogeny and eukaryogenesis. *Molecular Biology and Evolution, 39*(4). https://doi.org/10.1093/molbev/msac054.

55. Khanolkar, R. A., Clark, S. T., Wang, P. W., Hwang, D. M., Yau, Y. C. W., Waters, V. J., & Guttman, D. S. (2020). Ecological succession of polymicrobial communities in the cystic fibrosis airways. *mSystems, 5*(6). https://doi.org/10.1128/msystems.00809-20.

56. Kumar, K., Ghosh, S., Angelidaki, I., Holdt, S. L., Karakashev, D. B., Morales, M. A., & Das, D. (2016). Recent developments on biofuels production from microalgae and macroalgae. *Renewable and Sustainable Energy Reviews, 65*, 235–249. https://doi.org/10.1016/j.rser.2016.06.055.

57. Ławniczak, Ł., Woźniak-Karczewska, M., Loibner, A. P., Heipieper, H. J., & Chrzanowski, Ł. (2020). Microbial degradation of hydrocarbons—Basic principles for bioremediation: A review. *Molecules, 25*(4), 4. https://doi.org/10.3390/molecules25040856.

58. Lederberg, J., & Relman, D. A. (2009). *Microbial Evolution and Co-Adaptation: A Tribute to the Life and Scientific Legacies of Joshua Lederberg*: *Workshop Summary*. National Academies Press.

59. Lefkowitz, E. J., Dempsey, D. M., Hendrickson, R. C., Orton, R. J., Siddell, S. G., & Smith, D. B. (2018). Virus taxonomy: The database of the International Committee on Taxonomy of Viruses (ICTV). *Nucleic Acids Research, 46*(D1), D708–D717. https://doi.org/10.1093/nar/gkx932.

60. Lievens, B., Hallsworth, J. E., Pozo, M. I., Belgacem, Z. B., Stevenson,

A., Willems, K. A., & Jacquemyn, H. (2015). Microbiology of sugar-rich environments: Diversity, ecology and system constraints. *Environmental Microbiology, 17*(2), 278–298. https://doi.org/10.1111/1462-2920.12570.

61. Lorusso, A. (2020). Novel coronavirus (SARS-CoV-2) epidemic: A veterinary perspective. *Veterinaria Italiana, 1*. https://doi.org/10.12834/VetIt.2173.11599.1.

62. Lozupone, C. A., & Knight, R. (2007). Global patterns in bacterial diversity. *Proceedings of the National Academy of Sciences, 104*(27), 11436–11440. https://doi.org/10.1073/pnas.0611525104.

63. Madigan, M. T., Bender, K. S., Buckley, D. H., Sattley, W. M., & Stahl, D. A. (2018). *Brock Biology of Microorganisms* (15th ed.). Pearson.

64. Magouras, I., Brookes, V., Jori, F., Martin, A., Pfeiffer, D. U., & Dürr, S. (2020). Emerging zoonotic diseases: Should we rethink the animal–human interface? *Frontiers in Veterinary Science, 7*. https://doi.org/10.3389/fvets.2020.582743.

65. Maier, R. M., Pepper, I. L., & Gerba, C. P. (2009). *Environmental Microbiology* (2nd ed.). Academic Press.

66. Malik, J. A., & Goyal, M. R. (2022). *Bioremediation and Phytoremediation Technologies in Sustainable Soil Management: Inventive Techniques, Research Methods, and Case Studies* (1st ed.). Apple Academic Press. https://doi.org/10.1201/9781003281177.

67. Malik, J. A., Rahman, M. M., & Rais, N. (2022). Chapter 1–Microbial biotechnology: An introduction. In J. A. Malik (Ed.), *Microbes and Microbial Biotechnology for Green Remediation* (pp. 3–11). Elsevier. https://doi.org/10.1016/B978-0-323-90452-0.00035-9.

68. Mañas, P., & Pagán, R. (2005). Microbial inactivation by new technologies of food preservation. *Journal of Applied Microbiology, 98*(6), 1387–1399. https://doi.org/10.1111/j.1365-2672.2005.02561.x.

69. Manske, M., Miotto, O., Campino, S., Auburn, S., Almagro-Garcia, J., Maslen, G., O'Brien, J., Djimde, A., Doumbo, O., Zongo, I., Ouedraogo, J. B., Michon, P., Mueller, I., Siba, P., Nzila, A., Borrmann, S., Kiara, S. M., Marsh, K., Jiang, H., & Kwiatkowski, D. P. (2012). Analysis of *Plasmodium falciparum* diversity in natural infections by deep sequencing. *Nature, 487*(7407), 375–379. https://doi.org/10.1038/nature11174.

70. Martiny, J. B. H., Bohannan, B. J. M., Brown, J. H., Colwell, R. K., Fuhrman, J. A., Green, J. L., Horner-Devine, M. C., Kane, M., Krumins, J. A., Kuske, C. R., Morin, P. J., Naeem, S., Øvreås, L., Reysenbach, A. L., Smith, V. H., & Staley, J. T. (2006). Microbial biogeography: Putting microorganisms on the map. *Nature Reviews Microbiology, 4*(2), 102–112. https://doi.org/10.1038/nrmicro1341.

71. Matulich, K. L., Mohamed, D. J., & Martiny, J. B. H. (2013). Microbial biodiversity. In S. A. Levin (Ed.), *Encyclopedia of Biodiversity* (2nd ed., pp. 252–258). Academic Press. https://doi.org/10.1016/B978-0-12-384719-5.00236-7.
72. Mazzeo, L., & Piemonte, V. (2020). Fermentation and biochemical engineering: Principles and applications. In *Studies in Surface Science and Catalysis* (pp. 261–285). https://doi.org/10.1016/b978-0-444-64337-7.00015-x.
73. McCloskey, B., Dar, O., Zumla, A., & Heymann, D. L. (2014). Emerging infectious diseases and pandemic potential: Status quo and reducing risk of global spread. *The Lancet Infectious Diseases, 14*(10), 1001–1010. https://doi.org/10.1016/S1473-3099(14)70846-1.
74. Meechan, P. J., & Potts, J. (2020). Biosafety in microbiological and biomedical laboratories. https://stacks.cdc.gov/view/cdc/97733 (accessed on 9 November 2024).
75. Meena, M., Swapnil, P., Zehra, A., Aamir, M., Dubey, M. K., Patel, C. B., & Upadhyay, R. S. (2019). Virulence factors and their associated genes in microbes. In H. B. Singh, V. K. Gupta, & S. Jogaiah (Eds.), *New and Future Developments in Microbial Biotechnology and Bioengineering* (pp. 181–208). Elsevier. https://doi.org/10.1016/B978-0-444-63503-7.00011-5.
76. Merino, N., Aronson, H. S., Bojanova, D. P., Feyhl-Buska, J., Wong, M. L., Zhang, S., & Giovannelli, D. (2019). Living at the extremes: Extremophiles and the limits of life in a planetary context. *Frontiers in Microbiology, 10*. https://doi.org/10.3389/fmicb.2019.00780.
77. Milgroom, M. G. (2023). Protozoa. In *Springer eBooks* (pp. 71–87). https://doi.org/10.1007/978-3-031-38941-2_6.
78. Moat, A. G., Foster, J. W., & Spector, M. P. (2002). *Microbial Physiology*. John Wiley & Sons.
79. Morrison, L., & Zembower, T. R. (2020). Antimicrobial resistance. *Gastrointestinal Endoscopy Clinics, 30*(4), 619–635. https://doi.org/10.1016/j.giec.2020.06.004.
80. Morse, S. S. (2012). Public health surveillance and infectious disease detection. *Biosecurity and Bioterrorism: Biodefense Strategy, Practice, and Science, 10*(1), 6–16. https://doi.org/10.1089/bsp.2011.0088.
81. Mukherjee, S. (2023). Bacteria: Definition & characteristics with examples & diagram. *Science Facts*. https://www.sciencefacts.net/bacteria-cell.html (accessed on 9 November 2024).

82. Murray, P. R., Rosenthal, K. S., & Pfaller, M. A. (2015). *Medical Microbiology* (8th ed.). Elsevier Health Sciences.

83. Nagy, A., & Alhatlani, B. (2021). An overview of current COVID-19 vaccine platforms. *Computational and Structural Biotechnology Journal, 19*, 2508–2517. https://doi.org/10.1016/j.csbj.2021.04.061.

84. National Research Council, Affairs, P. and G., Cooperation, D., Security, and, & Biotechnology, C. on R. S. and P. to P. the D. A. of. (2004). *Biotechnology Research in an Age of Terrorism*. National Academies Press.

85. Nazemi, E., Dubowski, J. J., & Frost, E. (2017). *Nazemi, E., "Photonic Monitoring of Biological Activities of Bacteria Immobilized on Biofunctionalized...."* ResearchGate. https://www.researchgate.net/publication/327558675_Nazemi_E_PHOTONIC_ MONITORING_OF_BIOLOGICAL_ ACTIVITIES_OF_BACTERIA_IMMOBILIZED_ON_BIOFUNCTIONALIZED_ SURFACES_OF_QUANTUM_SEMICONDUCTORS_PhD_in_Electrical_ Engineering_2017_Universite_de_Sherbrooke?_tp=eyJjb250ZXh0Ijp7ImZpcnN0UGFnZ SI6Il9kaXJlY3QiLCJwYWdlIjoiX2RpcmVjdCJ9fQ (accessed on 9 Novemb er 2024).

86. Ni, G., Lappan, R., Hernández, M., Santini, T., Tomkins, A. G., & Greening, C. (2023). Functional basis of primary succession: Traits of the pioneer microbes. *Environmental Microbiology, 25*(1), 171–176. https://doi.org/10.1111/1462-2920.16266.

87. Nielsen, J., & Keasling, J. D. (2016). Engineering cellular metabolism. *Cell, 164*(6), 1185–1197. https://doi.org/10.1016/j.cell.2016.02.004.

88. O'Mathúna, D. P. (2007). Bioethics and biotechnology. *Cytotechnology, 53*(1–3), 113–119. https://doi.org/10.1007/s10616-007-9053-8.

89. Olempska-Beer, Z. S., Merker, R. I., Ditto, M. D., & DiNovi, M. J. (2006). Food-processing enzymes from recombinant microorganisms—A review. *Regulatory Toxicology and Pharmacology, 45*(2), 144–158. https://doi.org/10.1016/j.yrtph.2006.05.001.

90. Ōmura, S., Ikeda, H., Ishikawa, J., Hanamoto, A., Takahashi, C., Shinose, M., Takahashi, Y., Horikawa, H., Nakazawa, H., Osonoe, T., Kikuchi, H., Shiba, T., Sakaki, Y., & Hattori, M. (2001). Genome sequence of an industrial microorganism *Streptomyces avermitilis*: Deducing the ability of producing secondary metabolites. *Proceedings of the National Academy of Sciences, 98*(21), 12215–12220. https://doi.org/10.1073/pnas.211433198.

91. Pérez-Brocal, V., Gil, R., Ramos, S., Lamelas, A., Postigo, M., Michelena, J. M., Silva, F. J., Moya, A., & Latorre, A. (2006). A small microbial

genome: The end of a long symbiotic relationship? *Science, 314*(5797), 312–313. https://doi.org/10.1126/science.1130441.

92. Qiu, H., Yoon, H. S., & Bhattacharya, D. (2013). Algal endosymbionts as vectors of horizontal gene transfer in photosynthetic eukaryotes. *Frontiers in Plant Science, 4*. https://doi.org/10.3389/fpls.2013.00366.
93. Rajendran, P., & Gunasekaran, P. (2019). *Microbial Bioremediation*. MJP Publisher.
94. Rao, N. S. S. (2016). *Advances in Agricultural Microbiology*. Elsevier.
95. Read, D. J., & Perez-Moreno, J. (2003). Mycorrhizas and nutrient cycling in ecosystems – a journey towards relevance? *New Phytologist, 157*(3), 475–492. https://doi.org/10.1046/j.1469-8137.2003.00704.x.
96. Reddy, C. M., Arey, J. S., Seewald, J. S., Sylva, S. P., Lemkau, K. L., Nelson, R. K., Carmichael, C. A., McIntyre, C. P., Fenwick, J., Ventura, G. T., Van Mooy, B. A. S., & Camilli, R. (2012). Composition and fate of gas and oil released to the water column during the Deepwater Horizon oil spill. *Proceedings of the National Academy of Sciences of the United States of America, 109*(50), 20229–20234. https://doi.org/10.1073/pnas.1101242108.
97. Roller, B. R. K., Stoddard, S. F., & Schmidt, T. M. (2016). Exploiting rRNA operon copy number to investigate bacterial reproductive strategies. *Nature Microbiology, 1*(11), 1–7. https://doi.org/10.1038/nmicrobiol.2016.160.
98. Saeed, M. U., Hussain, N., Sumrin, A., Shahbaz, A., Noor, S., Bilal, M., Aleya, L., & Iqbal, H. M. N. (2022). Microbial bioremediation strategies with wastewater treatment potentialities–A review. *Science of The Total Environment, 818*, 151754. https://doi.org/10.1016/j.scitotenv.2021.151754.
99. Sales Da Silva, I. G., Gomes De Almeida, F. C., Padilha Da Rocha E Silva, N. M., Casazza, A. A., Converti, A., & Asfora Sarubbo, L. (2020). Soil bioremediation: Overview of technologies and trends. *Energies, 13*(18), 4664. https://doi.org/10.3390/en13184664.
100. Sammauria, R., Kumawat, S., Kumawat, P., Singh, J., & Jatwa, T. K. (2020). Microbial inoculants: Potential tool for sustainability of agricultural production systems. *Archives of Microbiology, 202*(4), 677–693. https://doi.org/10.1007/s00203-019-01795-w.
101. Singh, H. B., Jha, A., & Keswani, C. (2016). *Intellectual Property Issues in Biotechnology*. CABI.
102. Singh, R. (2019). Chapter 8–Microbial biotechnology: A promising implement for sustainable agriculture. In J. S. Singh & D. P. Singh (Eds.), *New and Future Developments in Microbial Biotechnology and*

Bioengineering (pp. 107–114). Elsevier. https://doi.org/10.1016/B978-0-444-64191-5.00008-0.

103. Sousa, A. M., Pereira, M. O., & Lourenço, A. (2015). MorphoCol: An ontology-based knowledgebase for the characterization of clinically significant bacterial colony morphologies. *Journal of Biomedical Informatics, 55*, 55–63. https://doi.org/10.1016/j.jbi.2015.03.007.
104. Stasi, A., & Thongpravati, O. (2024). Biotechnology innovation in Do-It-Yourself (DIY) gene editing: A call for a new regulatory framework. *OBM Genetics, 8*(2), 1–11. https://doi.org/10.21926/obm.genet.2402238.
105. Subhan, F., Shahzad, R., Tauseef, I., Haleem, K. S., Rehman, A. U., Mahmood, S., & Lee, I. J. (2018). Isolation, identification, and pathological effects of beach sand bacterial extract on human skin keratinocytes *in vitro*. *PeerJ, 6*, e4245. https://doi.org/10.7717/peerj.4245.
106. Tan, H. W., Pang, Y. L., Lim, S., & Chong, W. C. (2023). A state-of-the-art of phytoremediation approach for sustainable management of heavy metals recovery. *Environmental Technology & Innovation, 30*, 103043. https://doi.org/10.1016/j.eti.2023.103043.
107. Tay, W. H., Chong, K. K. L., & Kline, K. A. (2016). Polymicrobial–Host Interactions during Infection. *Journal of Molecular Biology, 428*(17), 3355–3371. https://doi.org/10.1016/j.jmb.2016.05.006.
108. Thammavongsa, V., Kim, H. K., Missiakas, D., & Schneewind, O. (2015). Staphylococcal manipulation of host immune responses. *Nature Reviews Microbiology, 13*(9), 529–543. https://doi.org/10.1038/nrmicro3521.
109. Tiedje, J. M., Bruns, M. A., Casadevall, A., Criddle, C. S., Eloe-Fadrosh, E., Karl, D. M., Nguyen, N. K., & Zhou, J. (2022). Microbes and climate change: A research prospectus for the future. *mBio, 13*(3), e00800-22. https://doi.org/10.1128/mbio.00800-22.
110. Timmis, K. N. (2002). Pseudomonas putida: A cosmopolitan opportunist par excellence. *Environmental Microbiology, 4*(12), 779–781. https://doi.org/10.1046/j.1462-2920.2002.00365.x.
111. Torsvik, V., & Øvreås, L. (2002). Microbial diversity and function in soil: From genes to ecosystems. *Current Opinion in Microbiology, 5*(3), 240–245. https://doi.org/10.1016/S1369-5274(02)00324-7.
112. Udaondo, Z., Molina, L., Segura, A., Duque, E., & Ramos, J. L. (2016). Analysis of the core genome and pangenome of *Pseudomonas putida*. *Environmental Microbiology, 18*(10), 3268–3283. https://doi.org/10.1111/1462-2920.13015.
113. van der Wal, A., Geydan, T. D., Kuyper, T. W., & de Boer, W. (2013). A thready affair: Linking fungal diversity and community dynamics to

terrestrial decomposition processes. *FEMS Microbiology Reviews, 37*(4), 477–494. https://doi.org/10.1111/1574-6976.12001.

114. Verma, S., & Kuila, A. (2019). Bioremediation of heavy metals by microbial process. *Environmental Technology & Innovation, 14*, 100369. https://doi.org/10.1016/j.eti.2019.100369.
115. Vincent, S. G. T., Jennerjahn, T., & Ramasamy, K. (2021). Chapter 6—Assessment of microbial structure and functions in coastal sediments. In S. G. T. Vincent, T. Jennerjahn, & K. Ramasamy (Eds.), *Microbial Communities in Coastal Sediments* (pp. 167–185). Elsevier. https://doi.org/10.1016/B978-0-12-815165-5.00006-6.
116. Vos, M., Wolf, A. B., Jennings, S. J., & Kowalchuk, G. A. (2013). Micro-scale determinants of bacterial diversity in soil. *FEMS Microbiology Reviews, 37*(6), 936–954. https://doi.org/10.1111/1574-6976.12023.
117. Vulin, C. (2014). *A Quantitative Approach to Microbial Population Growth Using Tailored Cylindrical Yeast Colonies. ResearchGate.* https://www.researchgate.net/publication/278827748_A_quantitative_approach_to_microbial_population_growth_using_tailored_cylindrical_yeast_colonies (accessed on 9 November 2024).
118. Wadhia, K. (2013). Microbial Assay for Risk Assessment (MARA). In *Springer eBooks* (pp. 699–708). https://doi.org/10.1007/978-94-007-5704-2_64.
119. Wagner, A., Whitaker, R. J., Krause, D. J., Heilers, J. H., Van Wolferen, M., Van Der Does, C., & Albers, S. V. (2017). Mechanisms of gene flow in archaea. *Nature Reviews Microbiology, 15*(8), 492–501. https://doi.org/10.1038/nrmicro.2017.41.
120. Wang, L., Zhang, X., Tang, C., Li, P., Zhu, R., Sun, J., Zhang, Y., Cui, H., Ma, J., Song, X., Zhang, W., Gao, X., Luo, X., You, L., Chen, Y., & Dai, Z. (2022). Engineering consortia by polymeric microbial swarmbots. *Nature Communications, 13*(1). https://doi.org/10.1038/s41467-022-31467-1.
121. Waoo, A. A., & Vyas, C. (2023). Chapter 16—Microbes for the remediation of heavy metal–contaminated soil. In R. T. Kapoor & M. P. Shah (Eds.), *Integrative Strategies for Bioremediation of Environmental Contaminants, Volume Two* (pp. 297–316). Academic Press. https://doi.org/10.1016/B978-0-443-14013-6.00014-7.
122. Wieser, A., Schneider, L., Jung, J., & Schubert, S. (2012). MALDI-TOF MS in microbiological diagnostics—Identification of microorganisms and beyond (mini review). *Applied Microbiology and Biotechnology, 93*(3), 965–974. https://doi.org/10.1007/s00253-011-3783-4.

123. Woodworth, M. H., Wang, T., Sitchenko, K. L., Carpentieri, C., Friedman-Moraco, R. J., & Kraft, C. S. (2017). Ethical considerations in microbial therapeutic clinical trials. *The New Bioethics: A Multidisciplinary Journal of Biotechnology and the Body, 23*(3), 210–218. https://doi.org/10.1080/20502877.2017.1387386.

124. World Health Organization. (2014). *Antimicrobial Resistance: Global Report on Surveillance*. World Health Organization. https://iris.who.int/handle/10665/112642 (accessed on 9 November 2024).

INDEX

C

D

E

F

G

H

R

S

T

U

V